# Cambridge Elements ☰

Elements in the Structure and Dynamics of Complex Networks
edited by
Guido Caldarelli
Ca' Foscari University of Venice

# THE SHORTEST PATH TO NETWORK GEOMETRY

## A Practical Guide to Basic Models and Applications

M. Ángeles Serrano
*University of Barcelona,
University of Barcelona
Institute of Complex
Systems (UBICS) and
Catalan Institution for
Research and Advanced
Studies (ICREA)*

Marián Boguñá
*University of Barcelona and
University of Barcelona
Institute of Complex
Systems (UBICS)*

CAMBRIDGE
UNIVERSITY PRESS

# CAMBRIDGE
## UNIVERSITY PRESS

University Printing House, Cambridge CB2 8BS, United Kingdom

One Liberty Plaza, 20th Floor, New York, NY 10006, USA

477 Williamstown Road, Port Melbourne, VIC 3207, Australia

314–321, 3rd Floor, Plot 3, Splendor Forum, Jasola District Centre,
New Delhi – 110025, India

103 Penang Road, #05–06/07, Visioncrest Commercial, Singapore 238467

Cambridge University Press is part of the University of Cambridge.

It furthers the University's mission by disseminating knowledge in the pursuit of
education, learning, and research at the highest international levels of excellence.

www.cambridge.org
Information on this title: www.cambridge.org/9781108791083
DOI: 10.1017/9781108865791

© M. Ángeles Serrano and Marián Boguñá 2021

First published 2021

A catalogue record for this publication is available from the British Library.

ISBN 978-1-108-79108-3 Paperback
ISSN 2516-5763 (online)
ISSN 2516-5755 (print)

# The Shortest Path to Network Geometry

## A Practical Guide to Basic Models and Applications

Elements in the Structure and Dynamics of Complex Networks

DOI: 10.1017/9781108865791
First published online: December 2021

M. Ángeles Serrano
*University of Barcelona, University of Barcelona Institute of Complex Systems (UBICS) and Catalan Institution for Research and Advanced Studies (ICREA)*
Marián Boguñá
*University of Barcelona and University of Barcelona Institute of Complex Systems (UBICS)*
**Author for correspondence:** M. Ángeles Serrano, marian.serrano@ub.edu

**Abstract:** Real networks comprise hundreds to millions of interacting elements and permeate all contexts, from technology to biology to society. All of them display non-trivial connectivity patterns, including the small-world phenomenon, making nodes to be separated by a small number of intermediate links. As a consequence, networks present an apparent lack of metric structure and are difficult to map. Yet, many networks have a hidden geometry that enables meaningful maps in the two-dimensional hyperbolic plane. The discovery of such hidden geometry and the understanding of its role have become fundamental questions in network science, giving rise to the field of network geometry. This Element reviews fundamental models and methods for the geometric description of real networks with a focus on applications of real network maps, including decentralized routing protocols, geometric community detection, and the self-similar multiscale unfolding of networks by geometric renormalization.

**Keywords:** network geometry, mapping techniques, navigability, community detection, renormalization

ISBNs: 9781108791083 (PB), 9781108865791 (OC)
ISSNs: 2516-5763 (online), 2516-5755 (print)

# Contents

# 1 From Networks to Maps

Throughout history, maps have been at the center of political, economic, and geostrategic decisions and have become a critical piece of our everyday lives, serving as a precise and relevant information source. Maps provide an accurate way of visualizing, storing, and communicating information, help to recognize locational distributions, spatial patterns, and relationships, and allow us to track processes that operate through space at different length scales. Our work in the last decade led us to prove that many real complex networks are natural geometric objects and can be mapped into hidden low-dimensional metric spaces with hyperbolic geometry, where distances determine the likelihood of the interactions and encode the different intrinsic attributes determining how similar the elements of the system are (Allard et al. [2017]; Boguñá, Papadopoulos, and Krioukov [2010]; García-Pérez et al. [2016]; García-Pérez, Boguñá, and Serrano [2018]; Kleineberg et al. [2016]; Krioukov et al. [2012]; Krioukov et al. [2010]; Papadopoulos et al. [2012]; Serrano, Boguñá, and Sagues [2012]; Serrano, Krioukov, and Boguñá [2008]). We took advantage of the large amount of empirical data available and the current explosion in computing power to create meaningful geometric maps of large real networks by embedding them in an underlying space that ought not to be geographical or spatially obvious. In this Element, we review our most relevant research on this topic, with a special focus on models and applications to real networks. These results triggered the field of network geometry to become one of the fundamental areas within network science devoted to the discovery and modeling of nontrivial geometric properties of complex networks (Boguñá et al. [2020]).

Complex networks typically have been studied as topological objects (Dorogovtsev and Mendes [2003]; M. E. J. Newman [2010]), graphs where elements are represented as nodes and their interactions as links. Graphs of real networks are not regular lattices nor are they completely disordered or random, and their structure is imprinted with universal features. One of the most paradigmatic examples is the small-world phenomenon, connecting every pair of nodes in a network, on average, by a small number of intermediate links (Amaral et al. [2000]; Watts and Strogatz [1998]). Other ubiquitous properties are scale-free, or heavy tailed, distributions of the number of connections per node (degree) (Barabási and Albert [1999]), with a few nodes linked to an enormous number of neighbors (hubs with very high degrees, while most other nodes are poorly connected), modularity, and hierarchical structure (Amaral [2008]). These and other prevalent features are not a mere curiosity but arise as the outcome of evolutionary pressures or functional needs and affect the dynamics that characterize or take place within and between networks (Barrat, Barthélemy, and Vespignani [2008]).

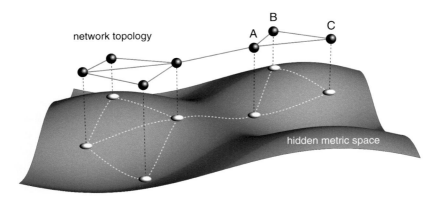

**Figure 1 Hidden metric spaces help to understand the structure and function of complex networks**. The smaller the distance between two nodes in the hidden metric space – the more similar they are – the more likely they are connected in the observable network topology. If node A is close to node B, and B is close to C, then A and C are necessarily close because of the triangle inequality in the metric space. Therefore, triangle ABC exists in the network topology with high probability, which explains the strong clustering observed in real complex networks.

One of the main consequences of the small-world effect is the apparent lack of a metric structure defined on the system. Certainly, in a small-world network, the distribution of shortest path lengths among pairs of nodes is sharply peaked around its average and, therefore, any pair of nodes is roughly separated by the same minimal number of intermediate links. This is the reason why complex networks are often considered as pure topological objects and are difficult to map. Yet, many networks are embedded in metric spaces. Some are explicit (Barthélemy [2011]) – like in airport networks (Barrat et al. [2004]; Guimerà et al. [2005]), power grids, or urban networks – whereas some are hidden yet shaping the network topological structure (Boguñá et al. [2010]; Krioukov et al. [2010]; Krioukov et al. [2009]; Serrano et al. [2008]); see Figure 1. This idea led to hidden metric space models of complex networks with an underlying effective hyperbolic geometry. These models are able to explain universal features observed in real-world systems, including the small-world property, scale-free degree distributions, clustering, and also fundamental mechanisms like preferential attachment in growing networks (Papadopoulos et al. [2012]), the emergence of communities (Zuev et al. [2015]), and multiscale self-similarity (García-Pérez, Boguñá, and Serrano [2018]). The discovery of the hidden geometry of real complex networks led to the emergence of the field of network geometry (Boguñá et al. [2020]), a major research area within network science.

The hidden metric space network models of complex networks couple their topology to an underlying geometry through a probabilistic connectivity law depending on distances in the space, which combine popularity and similarity dimensions in such a way that more popular and similar nodes have more chance to interact (Krioukov et al. [2010]; Papadopoulos et al. [2012]; Serrano et al. [2008]). The basic assumptions in our model are that there exists some similarity between nodes which, along degrees, plays an important role in how connections are established and that, since similarity is transitive, geometry is an appropriate mathematical formalism to encode it. The clue for the connection between topology and geometry is then clustering – transitive relationships, or triangles – which arises as a reflection in the topology of the network of the triangle inequality in the underlying hidden metric space. These models can be combined with statistical inference techniques to find the coordinates of the nodes in the underlying metric space that maximize the likelihood that the topology of the network is reproduced by the model (Blasius et al. [2018]; Boguñá et al. [2010]; García-Pérez et al. [2019]; Papadopoulos, Aldecoa, and Krioukov [2015]). One of the key properties of these maps is that the shortest paths in the topology of the networks follow closely geodesic lines in the underlying space. This ensures that networks highly congruent with the hidden metric space model are navigable, where navigability is understood as efficient routing of information based on the metric embedding (Allard and Serrano [2020]; Boguñá and Krioukov [2009]; Boguñá Krioukov, and Claffy [2009]; Boguñá et al. [2010]; Gulyás et al. [2015]; Krioukov et al. [2010]; Papadopoulos et al. [2010]).

One example of the power of this geometric approach is the discovery of the hyperbolic plane as the effective geometry of many real networks (see Fig. 2), including such disparate systems as the Internet at the Autonomous Systems level (Boguñá, Papadopoulos, and Krioukov [2010]), genome-scale reconstructions of metabolic networks in the cell (Serrano et al. [2012]), the World Trade Web from 1870 to 2013 (García-Pérez et al. [2016]), and brains of different species (Allard and Serrano [2020]). In the case of the Internet, the metric space provides a solution to the scalability limitations of current inter-domain routing protocols. For metabolic networks, it allows us to redefine the concept of biological pathways and to quantify their crosstalk. In international trade, the maps provide information about the long-term evolution of the system, unraveling the role of globalization, hierarchization, and localization forces. Finally, the effective geometry of human and nonhuman brain structures is also better described as hyperbolic than Euclidean, thus implying that hyperbolic embeddings are universal and meaningful maps of brain structure that allow for an efficient routing of information.

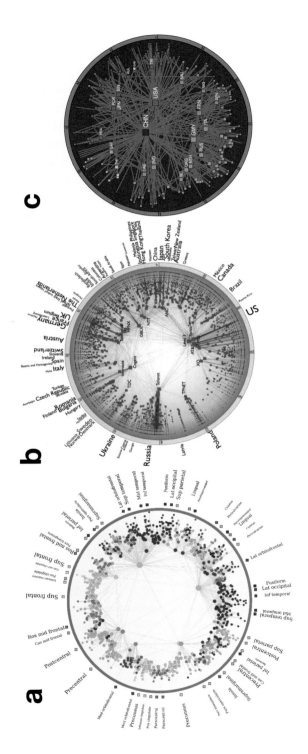

**Figure 2  Hidden metric space maps of real complex networks in the hyperbolic plane.** (a) Embedding of the human connectome, (b) Internet graph at the Autonomous Systems level, and (c) embedding of the World Trade Web as of year 2013. In all cases, the size of a node is proportional to the logarithm of its degree, and the radial coordinate decreases with increasing degree so that higher degree nodes are placed closer to the center of the disk. In (a) and (b) brain regions' and countries' names are located at the average angular position of all nodes belonging to the same region/country.

**Source:** Panel (a) modified from Zheng et al. (2020). Panel (b) reprinted from Boguñá et al. (2010). Panel (c) reprinted from García-Pérez et al. (2016).

These results suggest that the geometric paradigm improves our knowledge of the basic principles underlying the organization, function, and evolution of complex systems. But, in the long run, it also will transform research on how to model, predict, and control them, with potential implications for a large number of current challenges. These include efficient recommendation systems and search engines, prediction of epidemic spreading, and drug design in cancer and brain research.

## 2 Geometric Models for Static Topologies

Our first remarkable observation was to identify clustering – a measure of the number of triangles in a graph – as the key connection between complex networks and an underlying hidden geometry. Indeed, the triangle inequality in a metric space induces clustering in the structure of the graph, as illustrated in Figure 1. In Serrano et al. (2008), we analyzed the clustering coefficient of several real complex networks and found that their topological structure was compatible with an underlying hidden metric space. This finding led us to introduce the $\mathbb{S}^1$ class of network models (Serrano et al. [2008]). In these models, nodes are embedded in a metric space and connections exist with a gravity-law-like connection probability balancing the distance between nodes and their degrees; see Figure 3a. The connection probability encodes, in a simple and general way, the two major forces at play, namely, the effect of a similarity distance and the effect of the importance of the nodes involved in the connections. In this way, the model is able to generate scale-free, small-world, and clustered graphs very similar to those found in real complex networks, where the heterogeneity in the distribution of the number of contacts per node can be controlled independently of the level of clustering that measures the coupling with the metric space.

The $\mathbb{S}^1$ model is a mixed model in the sense that it combines a metric component and a topological component. Nodes are given coordinates in a metric similarity space but are also given degrees, determining their number of neighbors. At first glance, it seems difficult to combine, in a purely geometric framework, the small-world and scale-free properties that we observe in real networks. The major complication arises as a consequence of the small-world effect. This effect implies an exponential expansion of space, that is, the number of nodes within a disk of a certain radius grows exponentially with the radius (up to the finite size of the system). This behavior is in stark contrast to what happens in Euclidean spaces, where space grows as a power of the radius, but it is similar to what happens in hyperbolic geometry. In Krioukov et al. (2009, 2010), we developed the theory of random geometric graphs in hyperbolic geometry; see Figure 3b. Interestingly, scale-free graphs are the

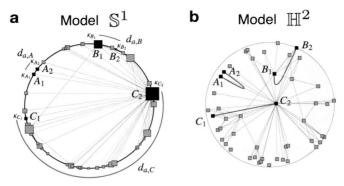

**Figure 3** (a) Model $\mathbb{S}^1$. The similarity distances $d_a$ between pairs of nodes A1-A2, B1-B2, and C1-C2 have been highlighted. The size of a node is proportional to its hidden degree $\kappa$. (b) Model $\mathbb{H}^2$ in the hyperbolic plane. Nodes in the different pairs are separated by the same hyperbolic distance. Nodes are equally sized, but nodes with higher hidden degree are positioned closer to the center. The similarity distance is the same in the two representations.

**Source:** Modified from the Supplementary Information in García-Pérez et al. [2016].

natural outcome of the formalism, indicating that this geometry is the most appropriate to model complex networks. Indeed, it produces in a natural way scale-free, small-world, and clustered graphs. However, the most surprising result is that this class of models, which we call $\mathbb{H}^2$, is isomorphic to our previous $\mathbb{S}^1$ version (Krioukov et al. [2010]; Serrano et al. [2008]).

This duality allows us to use either model indistinctly, depending on the particular application. The $\mathbb{S}^1$ version is especially convenient for theory development, analytical calculations, and the implementation of embedding techniques – which estimate the coordinates that maximize the likelihood of the observed structure being produced by the model. Instead, the $\mathbb{H}^2$ version is well suited for visualization purposes, to analyze navigation properties (Allard and Serrano [2020]; Boguñá and Krioukov [2009]; Boguñá et al. [2009]; Boguñá et al. [2010]; Gulyás et al. [2015]; Krioukov et al. [2010]; Papadopoulos et al. [2010]), or to define hierarchies within the network (García-Pérez et al. [2016]).

An interesting aspect is that our geometric class of models corresponds to an entropy-maximizing probabilistic mixture of grand canonical network ensembles, where network links can be thought of as noninteracting fermions whose energies depend on distances on the underlying geometry, with the particular choice of the functional form of this dependency defining network properties. At present, these models provide the simplest class of models capturing sparsity, the small-world property, power-law degree distributions, and nonvanishing clustering in a geometric framework with explicit symmetry structure (Boguñá et al. [2020]).

## 2.1 The $\mathbb{S}^1$ Model

In the $\mathbb{S}^1$ model (Serrano et al. [2008]), a node $i$ is assigned two hidden variables: a hidden degree $\kappa_i$ quantifying its popularity; and an angular position $\theta_i$ in a one-dimensional sphere (or circle), the similarity space, where distances with the other nodes serve as a proxy for their similarity. The radius of the circle is adjusted to $R = N/2\pi$, where $N$ is the number of nodes, so that the density is set to 1 without loss of generality. The probability of connection between any pair of nodes takes the form of a gravity law, whose magnitude increases with the product of the hidden degrees (i.e., their combined popularities) and decreases with the angular distance between the two nodes. In other words, more-similar nodes are angularly closer and are, therefore, more likely to be connected, whereas not-so-similar pairs of nodes have a high probability of being connected only if they are popular. Specifically, nodes $i$ and $j$ are connected with probability

$$p_{ij} = \frac{1}{1 + \chi_{ij}^{\beta}} = \frac{1}{1 + \left(\frac{d_{ij}}{\mu\kappa_i\kappa_j}\right)^{\beta}}, \tag{2.1}$$

where $\mu$ controls the average degree of the network, $\beta$ controls its level of clustering, and $d_{ij} = R\Delta\theta_{ij}$, and $\Delta\theta_{ij} = \pi - |\pi - |\theta_i - \theta_j||$ is the angular distance between nodes $i$ and $j$. Notice that there are no constraints on the distribution of hidden variables $\kappa$ and $\theta$. The angular distribution could be nonhomogeneous, and both hidden variables could even be correlated. This is an important observation because such angular inhomogeneities or correlations can explain the emergence of communities and other nontrivial topological patterns observed in real networks (Allard and Serrano [2020]).

A priori, the functional form of the connection probability could be any integrable function of argument $f\left(\frac{d_{ij}}{\mu\kappa_i\kappa_j}\right)$. However, the Fermi–Dirac form of the connection probability in Eq. (2.1) is the only possible choice that defines maximally random ensembles of graphs that are simultaneously sparse,[1] heterogeneous, clustered, small-worlds, and maximally degree–degree uncorrelated (Boguñá et al. [2020]).[2] Besides, with this choice, parameter $\beta$ has full control

---

[1] By sparse networks we mean ensembles of networks with size-independent average degree.

[2] By maximally degree–degree uncorrelated we mean that the probability of a node with hidden variable $\kappa$ having a neighbor with hidden variable $\kappa'$ is independent of $\kappa$. Yet, for heterogeneous scale-free networks, some level of degree–degree correlation is unavoidable, as shown in Boguñá, Pastor-Satorras, and Vespignani (2004).

of the level of clustering without affecting the degree distribution. It can be shown that the model undergoes a structural phase transition at $\beta = 1$ so that, for $\beta < 1$, networks are unclustered, whereas for $\beta > 1$, the ensemble generates networks with finite clustering in the thermodynamic limit (Serrano et al. [2008]).

---

### A SIMPLE ALGORITHM TO GENERATE NETWORKS FROM THE $\mathbb{S}^1$ ENSEMBLE

The algorithm below generates networks from the $\mathbb{S}^1$ ensemble in the limit $N \gg 1$, in the simple scenario of uncorrelated hidden variables $\kappa$ and $\theta$, and with the similarity coordinate homogeneously distributed.

1. Fix the number of nodes $N$, parameter $\beta > 1$, and the target average degree $\langle k \rangle$

2. Set $\mu$ to

$$\mu = \frac{\beta}{2\pi \langle k \rangle} \sin\left(\frac{\pi}{\beta}\right)$$

3. Assign a hidden degree $\kappa$ to every node from $\rho(\kappa)$ so that $\langle \kappa \rangle = \langle k \rangle$. Assign also an angular position $\theta$ to each node sampled uniformly at random within the interval $[0, 2\pi]$.

4. Connect every pair of nodes with probability given by Eq. (2.1).

---

With this parametrization – and in the thermodynamic limit – the expected degree of a node with hidden degree $\kappa$ is simply $\bar{k}(\kappa) = \kappa$, which justifies the name of hidden degree. Indeed, the expected degree of any node $i$ with hidden variables $(\kappa_i, \theta_i)$ can be evaluated as $\bar{k}(\kappa_i, \theta_i) = \sum_j p_{ij}$, where the connection probability is given in Eq. (2.1). If the network is homogeneous with respect to the similarity space, $\bar{k}(\kappa_i, \theta_i)$ is independent of $\theta_i$. Thus, the expected degree of any node with hidden degree $\kappa$, located without loss of generality at $\theta = 0$, can be expressed as

$$\bar{k}(\kappa) = 2\mu\kappa \int \kappa' \rho(\kappa') \left[ \int_0^{\frac{N}{2\mu\kappa\kappa'}} \frac{dt}{1 + t^\beta} \right] d\kappa' = \tag{2.2}$$

$$= N \int \rho(\kappa') {}_2F_1\left(1, \frac{1}{\beta}, 1 + \frac{1}{\beta}, -\left[\frac{N}{2\mu\kappa\kappa'}\right]^\beta\right) d\kappa',$$

where ${}_2F_1(1, \frac{1}{\beta}, 1 + \frac{1}{\beta}, -x^\beta)$ is the hypergeometric function, whose asymptotic behavior when $x \to \infty$ is ${}_2F_1(1, \frac{1}{\beta}, 1 + \frac{1}{\beta}, -x^\beta) \sim \pi \csc(\pi/\beta)/(\beta x)$.

Using this result, we recover the proportionality between expected and hidden degrees.

The degree distribution of the model can be evaluated as

$$P(k) = \frac{1}{k!} \int \kappa^k e^{-\kappa} \rho(\kappa) d\kappa, \tag{2.3}$$

that is, a mixture of Poisson distributions weighted by $\rho(\kappa)$. Eq. (2.3) shows that the model generates nodes with degree zero with probability $P(0) = \langle e^{-\kappa} \rangle$, so that the expected number on nonzero degree nodes is $N_{obs} = N[1 - P(0)]$, whereas the observable average degree (counting only nodes with nonzero degree) is $\langle k \rangle_{obs} = \langle k \rangle / [1 - P(0)]$.

In the case of scale-free networks, we consider $\rho(\kappa)$ to be a power-law distribution of the form

$$\rho(\kappa) = (\gamma - 1) \kappa_0^{\gamma-1} \kappa^{-\gamma} \; ; \; \kappa > \kappa_0 = \frac{\gamma - 2}{\gamma - 1} \langle k \rangle \; ; \; \gamma > 2. \tag{2.4}$$

Plugging this expression into Eq. (2.3), the degree distribution reads

$$P(k) = (\gamma - 1) \kappa_0^{\gamma-1} \frac{\Gamma(k + 1 - \gamma, \kappa_0)}{k!}, \tag{2.5}$$

where $\Gamma(k+1-\gamma, \kappa_0)$ is the incomplete gamma function, so that the asymptotic behavior of the degree distribution is $P(k) \sim k^{-\gamma}$. To simulate sparse scale-free networks with $\gamma < 2$ (as found, for instance, in airport networks) we need to introduce a cutoff in the distribution of hidden degrees $\kappa_c$. In particular, we choose a hard cutoff of the form

$$\rho(\kappa) = \frac{(\gamma - 1) \kappa_0^{\gamma-1}}{1 - \left(\frac{\kappa_c}{\kappa_0}\right)^{1-\gamma}} \kappa^{-\gamma} \text{ with } \kappa_0 < \kappa < \kappa_c, \tag{2.6}$$

where the lower cutoff is the solution of the equation

$$\langle k \rangle = \frac{\gamma - 1}{\gamma - 2} \kappa_0 \frac{1 - \left(\frac{\kappa_c}{\kappa_0}\right)^{2-\gamma}}{1 - \left(\frac{\kappa_c}{\kappa_0}\right)^{1-\gamma}}. \tag{2.7}$$

Equations (2.6) and (2.7) can also be used to compensate for finite size effects in scale-free networks with $\gamma \gtrsim 2$. Indeed, to prevent extreme fluctuations arising when $\gamma$ is very close to 2, instead of generating values of $\kappa$ from the unbounded distribution Eq. (2.4), we introduce a hard cutoff whose value is the same as the natural cutoff of the unbounded distribution, which can be approximated by $\kappa_c = \kappa_0 N^{1/(\gamma-1)}$ (Boguñá, Pastor-Satorras, and Vespignani [2004]). Then, we generate values of $\kappa$ from Eq. (2.6) with parameter $\kappa_0$ equal to

$$\kappa_0 = \frac{1 - N^{-1}}{1 - N^{\frac{2-\gamma}{\gamma-1}}} \frac{\gamma - 2}{\gamma - 1} \langle k \rangle. \tag{2.8}$$

Notice that when $\gamma$ is very close to 2, finite size effects can be very important even for large networks. However, notice that this is not the only source of finite size effects. To fully account for finite size effects, we must also consider the effects coming from the upper limit in the integral in Eq. (2.2), as done in García-Pérez et al. (2019). However, in many practical applications, the correction in Eq. (2.8) is enough.

The $\mathbb{S}^1$ model can be used to produce synthetic ensembles with controllable structural features or for high-fidelity replication of a specific real network. To that end, observed degrees in the real network can be taken as good proxies of hidden degrees, and parameters $\mu$ and $\beta$ can be tuned to reproduce the average degree and clustering of the real network. This procedure is not very accurate for heterogeneous networks due to strong fluctuations. Actual hidden degrees could be estimated from real data to avoid the mismatch between hidden and observed degrees, but this operation can be demanding and, besides, there is no guarantee that all nodes end up with the exact same degree they had in the real network. An alternative is the implementation of the geometric randomization model introduced in Starnini, Ortiz, and Serrano (2019), which preserves exactly the degree sequence of the input network while producing a version of the network maximally congruent with the $\mathbb{S}^1$ model.

The geometric randomization model assumes the same form of the connection probability as in the $\mathbb{S}^1$ model. Given a real network, nodes are given angular coordinates in the similarity space uniformly at random. Then, the network is rewired in order to maximize the likelihood that the new topology is generated by the $\mathbb{S}^1$ model while preserving the observed degrees and, thus, the total number of edges. After selecting a value of $\beta$, for instance, the one that replicates the level of clustering of the original network, the rewiring procedure is conducted by executing a Metropolis–Hastings algorithm as follows.

### Geometric Randomization Model

1. Assign each node an angular coordinate uniformly at random.
2. Choose two links at random, say between nodes $i$ and $j$ and between nodes $l$ and $m$.
3. Compute the probability of rewiring (connecting $i$ and $l$ and $j$ and $m$) as

$$p_r = \min\left[1, \frac{\mathcal{L}_{new}}{\mathcal{L}_{old}}\right] = \min\left[1, \left(\frac{\Delta\theta_{ij}\Delta\theta_{lm}}{\Delta\theta_{il}\Delta\theta_{jm}}\right)^\beta\right], \qquad (2.9)$$

where $\mathcal{L}_{new}$ corresponds to the value of the likelihood function after the swap and $\mathcal{L}_{old}$ before the swap, both evaluated using Eq. (3.1) (see next section) and the probability of connection in Eq. (2.1). Notice that $p_r$ only requires information about the angular coordinates of nodes.

4. Accept the link swap with probability $p_r$, making sure not to create a multiple link or a self-loop.
5. Stop the algorithm after the likelihood fluctuates steadily around a plateau.

Notice that the resulting network might not have global connectivity even if the degrees of the nodes have not changed. The geometric randomization model can be used as a null model in the analysis of features which are particularly sensitive to fluctuations in the degree cutoff, like the behavior of dynamical processes such as epidemic spreading or synchronization.

## 2.2 The $\mathbb{H}^2$ Model

The $\mathbb{H}^2$ model (Krioukov et al. [2010, 2009]) is a quasi-isomorphic purely geometric version of the $\mathbb{S}^1$ model, in which the popularity and similarity dimensions are combined into a single distance in the hyperbolic plane, such that hyperbolically closer nodes are more likely to be connected; see Figure 3b.

There are many representations of the hyperbolic plane $\mathbb{H}^2$. In the two-dimensional hyperboloid model that we use in our work, points in $\mathbb{H}^2$ (of constant curvature $K = -1$) are characterized by two coordinates $(r, \theta)$ and the metric tensor reads

$$ds^2 = dr^2 + \sinh^2 r d\theta^2. \tag{2.10}$$

By comparing it to the metric of the familiar Euclidean two-dimensional plane in polar coordinates, $ds^2 = dr^2 + r^2 d\theta^2$, we immediately notice that the perimeters of hyperbolic circles and the areas of disks of radius $r$ grow much faster with $r$ than do those of Euclidean circles. In particular, they grow exponentially fast when $r \gg 1$, akin to growth behavior in trees. This is the first indication that networks in hyperbolic geometry may be endowed with the small-world property while still being purely geometric.

The hyperbolic distance between two points at radial coordinates $r_i$ and $r_j$, separated by an angular distance $\Delta\theta_{ij}$, can be computed with the hyperbolic law of cosines and reads

$$\cosh x_{ij} = \cosh r_i \cosh r_j - \sinh r_i \sinh r_j \cos \Delta\theta_{ij}. \tag{2.11}$$

This expression can be rewritten in a more convenient way:

$$\cosh x_{ij} = \cosh(r_i - r_j)\left[1 - \sin^2 \frac{\Delta\theta_{ij}}{2}\right] + \cosh(r_i + r_j) \sin^2 \frac{\Delta\theta_{ij}}{2}. \tag{2.12}$$

From the previous equation, it is easy to see that when

$$\sin^2 \frac{\Delta\theta_{ij}}{2} \gg \frac{\cosh(r_i - r_j)}{\cosh(r_i - r_j) + \cosh(r_i + r_j)}, \tag{2.13}$$

then the first term in the right side of Eq. (2.12) can be neglected. Besides, if $r_i > r_j \gg 1$, then the term in the right hand side of Eq. (2.13) goes as $e^{-2r_j}$ and the inequality Eq. (2.13) is fulfilled even for very small angular separations. Therefore, in this limit, a very good approximation for the hyperbolic distance is given by

$$x_{ij} \approx r_i + r_j + 2\ln\sin\frac{\Delta\theta_{ij}}{2} \approx r_i + r_j + 2\ln\frac{\Delta\theta_{ij}}{2}. \tag{2.14}$$

This expression will show its importance when analyzing the equivalence between the hyperbolic network model described next and the $\mathbb{S}^1$ model.

Networks in the $\mathbb{H}^2$ model are generated by distributing points within a disk of radius $R_{\mathbb{H}^2}$ in the hyperbolic plane and connecting pairs of nodes $i$ and $j$ with probability $p_{ij} = f(x_{ij} - R_{\mathbb{H}^2})$, where $x_{ij}$ is the hyperbolic distance between the two nodes and $f(\cdot)$ is a decreasing function of its argument. The distribution of points within the disk is, a priori, arbitrary and, as in the case of the $\mathbb{S}^1$ model, the radial and angular coordinates could be correlated. In the simplest case, the angular distribution is uniform and the radial distribution is proportional to the volume element of $\mathbb{H}^2$, so that points are then homogeneously distributed within the disk. This corresponds to distributing points radially according to the probability density

$$\rho(r) = \frac{\sinh r}{\cosh R_{\mathbb{H}^2} - 1}, \quad \text{with } r \in [0, R_{\mathbb{H}^2}]. \tag{2.15}$$

The model generates sparse graphs when $N \propto e^{\frac{R_{\mathbb{H}^2}}{2}}$, in which case the degree distribution has a power-law tail with exponent $\gamma = 3$ (Krioukov et al. [2010, 2009]). If, instead, nodes are distributed (quasi-uniformly) with the probability density

$$\rho(r) = \alpha\frac{\sinh \alpha r}{\cosh \alpha R_{\mathbb{H}^2} - 1}, \quad \text{with } r \in [0, R_{\mathbb{H}^2}] \tag{2.16}$$

and $\alpha \geq 1/2$, then the degree distribution has a power-law tail with exponent $\gamma = 2\alpha + 1$. In both cases, the underlying metric structure induces the emergence of clustering in the thermodynamic limit, modulated by the specific functional form of the connection probability $p_{ij}$. As in the $\mathbb{S}^1$ model, maximum entropy ensembles are obtained when this connection probability takes the Fermi–Dirac form,

$$p_{ij} = \frac{1}{1 + e^{\frac{\beta}{2}(x_{ij} - R_{\mathbb{H}^2})}}, \tag{2.17}$$

with $\beta > 1$. As in the $\mathbb{S}^1$ model, parameter $\beta$ controls the level of clustering of the ensemble, approaching zero when $\beta \to 1^+$ and converging to a constant value when $\beta \to \infty$. Interestingly, Eq. (2.17) suggests that we can interpret the network ensemble as a system of noninteracting fermions – the edges – that can occupy different available states – all possible pairs among nodes – with energies defined by the corresponding hyperbolic distances, $\beta$ being the inverse of the system's temperature, and $R_{\mathbb{H}^2}$ the chemical potential controlling the expected number of fermions (Krioukov et al. [2010]).

To establish a mapping between the $\mathbb{S}^1$ model and the $\mathbb{H}^2$ model, the angular coordinates remain as in the $\mathbb{S}^1$ model, but the hidden degrees are transformed into radial coordinates according to

$$r_i = R_{\mathbb{H}^2} - 2\ln\frac{\kappa_i}{\kappa_0}, \qquad (2.18)$$

where the radius of the two-dimensional hyperbolic disk containing all nodes is

$$R_{\mathbb{H}^2} = 2\ln\frac{N}{\pi\mu\kappa_0^2}. \qquad (2.19)$$

Higher-degree nodes are therefore located closer to the center of the $\mathbb{H}^2$ disk, whereas low-degree nodes are placed near its boundary. In fact, if hidden degrees are power-law distributed according to the probability density in Eq. (2.4), then the mapping Eq. (2.18) causes the radial coordinates to be distributed as in Eq. (2.16) for large $R_{\mathbb{H}^2}$.

Substituting Eqs. (2.18) and (2.19) into the connection probability of the $\mathbb{S}^1$ model Eq. (2.1) yields

$$p_{ij} = \frac{1}{1 + e^{\frac{\beta}{2}(\tilde{x}_{ij} - R_{\mathbb{H}^2})}}, \qquad (2.20)$$

where $\tilde{x}_{ij} = r_i + r_j + 2\ln\frac{\Delta\theta_{ij}}{2}$ is, as discussed above, a very good approximation of the hyperbolic distance between two points with coordinates $(r_i, \theta_i)$ and $(r_j, \theta_j)$ in the hyperbolic disk with curvature $K = -1$.[3] Besides, the discrepancy between $\tilde{x}_{ij}$ and the true hyperbolic distance $x_{ij}$ is not relevant in the case of networks. Indeed, the expected value of the smallest radial coordinate corresponds to the expected value of the largest hidden degree, which scales

---

[3] Note that the Fermi–Dirac connection probability is not a requirement for the mapping to hold. In fact, if the connection probability in the $\mathbb{S}^1$ model is an integrable but otherwise arbitrary function of $\frac{d_{ij}}{\mu\kappa_i\kappa_j}$, then in the hyperbolic representation the connection probability is a function of the argument $e^{\frac{1}{2}(\tilde{x}_{ij} - R_{\mathbb{H}^2})}$.

as $\kappa_c \sim \kappa_0 N^{\frac{1}{\gamma-1}}$, and so, it scales as $r_{min} \sim 2\frac{\gamma-2}{\gamma-1}\ln N$. For such nodes, the inequality in Eq. (2.13) becomes

$$\Delta\theta \gg 2N^{2\frac{2-\gamma}{\gamma-1}}. \tag{2.21}$$

However, in a homogeneous angular distribution, the average distance between two consecutive nodes – and so the minimal distance – is of the order $N^{-1}$, which is larger than the value in Eq. (2.21) for $\gamma > 3$ and, thus, the approximation $x_{ij} \approx \tilde{x}_{ij}$ holds for all pairs of nodes. When $2 < \gamma < 3$, the number of closest nodes to the hub for which the approximation does not hold exactly scales as $N^{\frac{3-\gamma}{\gamma-1}}$, which is a vanishing fraction of its total number of neighbors. Consequently, the number of pairs of nodes for which the approximation $x_{ij} \approx \tilde{x}_{ij}$ does not hold is a vanishing fraction in the thermodynamic limit. Besides, for all these pairs of nodes, the connection probability is very close to 1 both in the $\mathbb{S}^1$ and $\mathbb{H}^2$ models, so that those pairs of nodes are almost surely connected in both models. After these considerations, we conclude that the mapping between $\mathbb{S}^1$ and $\mathbb{H}^2$ models becomes exact in the thermodynamic limit. This mapping allows us to use both models indifferently depending on the application at hand.

A SIMPLE ALGORITHM TO GENERATE NETWORKS FROM THE $\mathbb{H}^2$ ENSEMBLE
The algorithm below generates scale-free networks with exponent $\gamma$ from the $\mathbb{H}^2$ ensemble in the limit $N \gg 1$.

1. Fix the number of nodes $N$, parameter $\beta > 1$, and the target average degree $\langle k \rangle$.
2. Set the radius of the hyperbolic disk to

$$R_{\mathbb{H}^2} = 2\ln\left[\frac{2N}{\beta\sin(\pi/\beta)\langle k\rangle}\left(\frac{\gamma-1}{\gamma-2}\right)^2\right].$$

3. Assign a radial coordinate $r$ to every node from Eq. (2.16). Assign also an angular position $\theta$ to each node sampled uniformly at random within the interval $[0, 2\pi]$.
4. Connect every pair of nodes with probability given by Eq. (2.17).

Alternatively, synthetic $\mathbb{H}^2$ networks can be obtained by generating synthetic $\mathbb{S}^1$ networks and transforming the obtained hidden degrees into radial coordinates using Eq. (2.18).

The two quasi-isomorphic models $\mathbb{S}^1$ and $\mathbb{H}^2$ generalize to spheres $\mathbb{S}^D$ of any dimension D and to hyperbolic spaces $\mathbb{H}^{D+1}$, with the connection probability being a function of the argument $\frac{d_{ij}}{(\mu\kappa_i\kappa_j)^{1/D}}$ in the case of the $\mathbb{S}^D$ model (Serrano et al. [2008]). Interestingly, since the group of symmetries of hyperbolic spaces is the Lorentz group SO(1,D+1), the equivalence between the two models is a reflection of the isomorphisms between the Lorentz group SO(1,D+1) and the Möbius group acting on the sphere $\mathbb{S}^D$ as the group of its conformal transformations. This isomorphism is a starting point of the anti-de Sitter/conformal field theory (AdS/CFT) correspondence in string theory (Maldacena [1998]).

Apart from the information in the previous paragraph, in this Element we present our hidden metric space network models in a two-dimensional underlying geometry, where a one-dimensional similarity coordinate is combined with a second popularity coordinate related to the degrees of the nodes. Even if our models in similarity dimension $D = 1$ are very good at reproducing the structure of real complex networks, there is no fundamental reason to believe that the similarity space should be one-dimensional for them all. For instance, the likelihood to trade between countries in the world can be dictated by cultural, political, geographical, administrative, and economic dimensions, and two countries can be at the same time close along one of them and far apart along another.

In fact, if networks are metric and related to an underlying space, the differences in the structure of that space due to changes in its dimensionality should be reflected in the topology of the resulting networks. In particular, it has been shown that the maximum clustering coefficient that can be obtained from a geometric model decreases as the dimension of the space increases; see supplementary material in García-Pérez, Boguñá, and Serrano (2018). This suggests that real-world networks, which typically exhibit strong clustering, must belong to the low-dimensionality regime. However, an excessive dimensionality reduction applied when mapping real networks into the underlying space would distort embedding distances, such that in general it would be difficult to tally the probability of connection in the networks with its metricity as measured by the clustering coefficient. Taken together, this indicates that the dimension of the similarity subspace of real complex networks will be typically small. However, in general, we will need more resolution than that provided by a single similarity coordinate to understand the different relationships that shape the structure of the network and to be able to reproduce its structural features with high fidelity.

## 3 Mapping Real Networks

Maps of real networks can be found by reverse-engineering our geometric models with statistical inference techniques to find the coordinates of the nodes in the underlying space that maximize the likelihood that the topology of the network is generated by the model, an approach that was implemented in different embedding tools (Blasius et al. [2018]; Boguñá et al. [2010]; Papadopoulos, Aldecoa, and Krioukov [2015]; Papadopoulos, Psomas, and Krioukov [2015]). More precisely, the maps are inferred by finding the hidden degree and angular position of each node, $\{(\kappa_i, \theta_i),\ i = 1, \cdots, N\}$, that maximize the likelihood that the structure of the network was generated by the $\mathbb{S}^1/\mathbb{H}^2$ model, where the likelihood $\mathcal{L}$ is evaluated as

$$\mathcal{L} = \prod_{i<j} \left[p_{ij}\right]^{a_{ij}} \left[1 - p_{ij}\right]^{1-a_{ij}} . \tag{3.1}$$

Here $p_{ij}$ is the connection probability, in our case given by Eq. (2.1), and $\{a_{ij}\}$ are the entries of the adjacency matrix of the network ($a_{ij} = a_{ji} = 1$ if nodes $i$ and $j$ are connected, $a_{ij} = a_{ji} = 0$ otherwise). This maximization, however, is computationally expensive and cannot be performed using a brute-force approach. To find meaningful results, we use heuristic optimization techniques that explore the fundamental properties of the model, for instance, the fact that, for scale-free networks, subgraphs of high-degree nodes have a higher internal average degree as compared to the complete network (Serrano et al. [2008]; Serrano, Krioukov, and Boguñá [2011]). This allows us to define an onion-like structure that helps us to guide the maximization process (Boguñá et al. [2010]; García-Pérez et al. [2019]; Papadopoulos, Aldecoa, and Krioukov [2015]; Papadopoulos, Psomas, and Krioukov [2015]).

Recently, unsupervised machine learning state-of-the-art techniques have also been used to embed complex networks into the hyperbolic plane, having a very competitive computational complexity (Alanis-Lobato, Mier, and Andrade-Navarro [2016a, 2016b]; Muscoloni et al. [2017]). These techniques project high-dimensional data to a much lower-dimensional space (Sarveniazi [2014]), leveraging on the redundancy of high-dimensional datasets where the original features are correlated, such that only a small subspace of the original representation space is eventually populated by the underlying process. Low-dimensional representations can then be produced with minimal information loss using dimensional reduction methods. However, the inherent randomness of real complex systems can degrade the quality of unsupervised methods that, on the other hand, will never inform about fundamental principles that explain the structure of the observed data. In our recently published embedding tool Mercator (García-Pérez et al. [2019]), we used the best of both machine learning data-driven and maximum likelihood model-driven approaches to find high-quality two-dimensional maps at the cost of an acceptable computational complexity.

Using these mapping tools, meaningful and reliable hyperbolic maps have been obtained for many real networks (Allard and Serrano [2020]; Boguñá et al. [2010]; García-Pérez et al. [2019]; García-Pérez et al. [2016]; Serrano et al. [2012]); see examples in Figure 2. The maps are highly congruent with metadata not contained in the topology of the network or provided in the embedding, like geopolitical information in the Internet or the World Trade Web (Boguñá et al. [2010]; García-Pérez et al. [2016]), biochemical pathways in metabolic networks (Serrano et al. [2012]), or neuroanatomical modules in the brain (Allard and Serrano [2020]). One of the key properties of these maps is that the shortest paths in the topology of real networks follow closely geodesic lines in hyperbolic space. In other words, real network maps are navigable. Navigability is understood as efficient transport of information, energy, matter, or other media based on the metric embedding without the global knowledge of the network structure and without finding shortest paths in the network, a computationally intensive combinatorial problem. Instead, latent space guides navigation in the network based on distances between nodes in the latent space. However, not all networks are intrinsically navigable; a combination of degree heterogeneity and clustering is needed to guarantee that geometric navigation is able to discover long-range links to approach the target node in few hops and, at the same time, to be able to find the target when reaching the local neighborhood of the target. Interestingly, the vast majority of real complex networks fulfill such constraints due to their scale-freeness and metric structure, which suggests the interesting possibility that some real complex networks evolved to optimize navigability (Boguñá et al. [2009]).

## 4 Mesoscale Organisation and Community Detection

One of the interesting empirical observations from these maps is that, in real networks, nodes are not distributed homogeneously in the hyperbolic plane. They display a radial gradient related to the degree distribution, so that higher-degree nodes tend to appear closer to the center of the disk. More importantly, nodes clusterize in the similarity subspace into densely populated angular regions separated by voids zones, which correlates well with metadata not explicitly contained in the topology of the network. For instance, in the case of the Internet, Autonomous Systems in the same country tend to be close in angular distance (Boguñá et al. [2010]). In the World Trade Web (García-Pérez et al. [2016]), our mapping is highly congruent with geopolitical aspects, placing in the same angular sectors countries that are either geographically or politically close. The same type of results can be observed for metabolic networks in relation to biological pathways (Serrano et al. [2012]) and for the brain and neuroanatomical modules (Allard and Serrano [2020]).

This observation poses fundamental questions about the origin of such communities and, at the same time, it offers the opportunity to design completely new methods to detect them in real networks. While communities can be set by hand in our models by prescribing a given angular distribution of nodes (Muscoloni and Cannistraci [2018]), in García-Pérez, Serrano, and Boguñá (2018) and Zuev et al. (2015) we introduced a new mechanism that *explains* their emergence. The mechanism is a generalization of the concept of preferential attachment called geometric preferential attachment and tries to mimic the fact that newborn nodes have a preference to appear in similarity areas that are already highly populated. A single parameter, $\Lambda$, modulates the strength of the mechanism interpolating between a purely random angular choice for newborn nodes and a strong preference to emerge close to already highly populated regions. In this way, it is possible to generate random geometric networks with the same local properties as the original models but with a complex mesoscale organization, as found in real networks.

Beyond these theoretic considerations, the angular inhomogeneities observed in maps of real networks suggest new methods to detect their community organization. The main idea is to use the angular distribution of nodes in the similarity subspace to cluster nodes into what we call soft communities. This clusterization can be performed in different ways. For instance, in our first method, named *Geometric Critical Gap Method* (G-CGM) (Serrano et al. [2012]; Zuev et al. [2015]), angular gaps, $\Delta\theta$, between consecutive nodes are measured, and values that exceed a certain critical value, $\Delta\theta_c$, are considered to separate adjacent soft communities of nodes in the similarity circle, which determines a unique partition into nonoverlapping communities. The critical gap is defined as the expected value of the largest gap that would be obtained if the $N$ nodes were distributed uniformly at random in $[0, 2\pi]$, so that no communities are expected in this case. Following this idea, the critical gap can be calculated assuming a Poisson point process on the circle of unit radius with density $\delta = N/(2\pi)$. In this case, the distribution of the angular gaps is approximately exponential with rate $\delta$, and the critical gap is given by

$$\Delta\theta_c = 2\pi \frac{\ln N + \gamma}{N}, \tag{4.1}$$

where $\gamma$ is the Euler–Mascheroni constant (Zuev et al. [2015]).

Our second method, the *Topological Critical Gap Method* (T-CGM) (García-Pérez et al. [2016]), is a hybrid method combining geometry and topology to select the partition of nodes in similarity space that maximizes modularity $Q$, where modularity is the standard metric quantifying the quality of the division of networks into clusters (Newman and Girvan [2004]). The method is implemented as follows.

> ### TOPOLOGICAL CRITICAL GAP METHOD
> Given a map of a real network:
>
> 1. Measure and sort all angular gaps from the smallest to the largest, $\{\Delta\theta_i; i = 1, N\}$ with $\Delta\theta_1 < \Delta\theta_2 < \cdots < \Delta\theta_N$.
> 2. Set the critical gap $\Delta\theta_c = \Delta\theta_1$ so that each node is in its own community. Compute the modularity $Q_1$ and set $Q_{max} = Q_1$.
> 3. Set the critical gap $\Delta\theta_c = \Delta\theta_2$, find the corresponding partition and compute its modularity $Q_2$. If $Q_2 > Q_{max}$ set $Q_{max} = Q_2$.
> 4. Repeat the previous step for all values of $\Delta\theta_i$ and update, if needed, $Q_{max}$.
> 5. Select the partition with the final $Q_{max}$.

At the end of the process, the partition corresponding to $Q_{max}$ is the optimal one delivered by the algorithm. Interestingly, the mutual information between partitions found by our method and by the Louvain algorithm (Blondel et al. [2008]) show that both partitions have a very high overlap, making the T-CGM an alternative method to detect communities (García-Pérez et al. [2016]). However, the number of communities discovered by T-CGM and the corresponding modularity is typically higher than that of the Louvain method, as, in general, Louvain modules integrate smaller T-CGM communities. From this point of view, we could say that the T-CGM has a better resolution than the Louvain method.

## 5 Self-Similarity and Renormalisation

Self-similarity has been studied in real complex networks from different perspectives (Alvarez-Hamelin et al. [2008]; García-Pérez, Boguñá, and Serrano [2018]; Kim et al. [2007]; Serrano et al. [2008, 2011]; Song, Havlin, and Makse [2005]). The first obvious observation is that in many networks, the degree distribution can be approximated by a power law, a first indication of a self-similar organization. A first step to unravel self-similarity was made in Song et al. (2005), where a renormalization scheme was defined based on topological shortest paths. However, shortest paths are a poor source of metric distances due to the small-world effect, and results were limited. The geometric approach provides an alternative that has unraveled self-similarity in networks from the analysis of the topological properties of nested hierarchies of subgraphs (Serrano et al. [2008, 2011]), and from a geometric renormalization technique applied to network maps discovered by our algorithms (García-Pérez, Boguñá, and Serrano [2018]). Next, we discuss the two symmetries.

## 5.1 Self-Similarity of Nested Subgraphs

Given a real network, there is an arbitrary number of ways one can define a hierarchy of nested subgraphs. A priori, given one such hierarchy, we cannot expect their subgraphs to have the same topological properties as in the original network. However, in real-world scale-free networks, the hierarchy defined by selecting nodes with degree above a given threshold $k_T$ defines a sequence of graphs that, except for a larger internal average degree, have the very same topological properties when rescaled by the internal average degree. Figure 4 shows a scheme of the filtering procedure and examples of the behavior of the clustering coefficient for the Internet and PGP graphs (Serrano et al. [2008]). Even though the internal average degree increases almost a decade from $k_T = 0$ to $k_T = 100$ in both networks, the behavior of the clustering coefficient remains invariant, except for finite size fluctuations. The figure also shows the same procedure applied to randomized versions of the same networks preserving the degree sequence. In this case, the self-similarity of the corresponding subgraphs is lost; see Serrano et al. (2008) for further details.

Interestingly, the $\mathbb{S}^1/\mathbb{H}^2$ model has this self-similarity property built in when hidden degrees are power law distributed and subgraphs are defined by selecting nodes with hidden degrees above a certain threshold $\kappa_T$, which, given the proportionality between $\kappa$ and expected degree, is equivalent to the procedure described above. In the thermodynamic limit, any (infinite size) such subgraph of a graph generated by the $\mathbb{S}^1$ model with average degree $\langle k \rangle$, exponent $\gamma$ and inverse temperature $\beta$ is a realization of the same $\mathbb{S}^1$ model with the same $\gamma$ and $\beta$ and an average degree given by

$$\langle k(\kappa_T) \rangle = \kappa_T^{3-\gamma} \langle k \rangle. \tag{5.1}$$

This result predicts an increasing internal average degree of subgraphs as we go deep in the hierarchy, exactly as we observe in real networks. Thus, the $\mathbb{S}^1/\mathbb{H}^2$ model explains the observed self-similarity of real networks and the behavior of the average degrees of self-similar subgraphs.[4]

---

[4] The same property is also present in the configuration model and in growing models of networks. However, in the former case, clustering vanishes in the thermodynamic limit, and in the latter

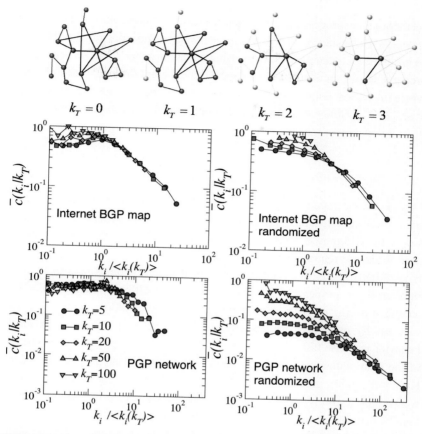

**Figure 4 Hierarchy of nested subgraphs.** Top row, a hierarchy of subgraphs is induced by removing nodes of degree below $k_T = 1, 2$, and 3 in the original graph. Bottom plots show the behavior of the clustering coefficient as a function of the rescaled degree for the Internet and PGP networks (left) and their randomized counterparts (right) for different values of the threshold degree $k_T$.

**Source:** Reprinted from Serrano et al. (2008).

This observation has a very important implication. Indeed, networks with a self-similar nested hierarchy of subgraphs and increasing internal average degree contain subgraphs that are macroscopic in the thermodynamic limit and have an arbitrary large average degree. This immediately affects the critical

---

the internal average degree is kept constant by construction, at odds with what is observed in real networks (Serrano et al. [2011]).

behavior of any process featuring a phase transition taking place on this type of networks. For instance, it predicts zero percolation and epidemic thresholds or an infinite critical temperature in the Ising model. In general, this self-similarity property sets either to zero or infinity the critical point for any phase transition where the critical point is a monotonic function of the average degree (Serrano et al. [2011]). Notice that this result is independent of the fact that the degree distribution may or may not be power law distributed.

## 5.2  Renormalization and Self-Similar Multiscale Unfoldings

Our geometric models enable a rigorous definition of self-similarity and scale invariance (Mandelbrot [1961]; Stanley [1971]) in real complex networks by affording a valid source of geometric length scales that can be used to design renormalization techniques (García-Pérez, Boguñá, and Serrano [2018]). In statistical physics, renormalization methods help to explore rigorously the properties of physical systems at different length scales by recursive averaging over short-distance degrees of freedom. This approach successfully explained, for instance, the universality properties of critical behavior in phase transitions (Wilson [1975, 1983]). Inspired by a precursor of the renormalization group, the block spin renormalization method of Leo Kadanoff (2000), previous efforts to understand the scaling behavior of networks took a purely topological approach and were based on shortest path lengths between nodes (Goh et al. [2006]; Kim et al. [2007]; Radicchi et al. [2008]; Rozenfeld, Song, and Makse [2010]; Song et al.[2005]; Havlin, and Makse [2006]). However, the collection of shortest paths is a poor source of length-based scaling factors in networks due to the small-world (Watts and Strogatz [1998]) or even ultrasmall-world (Cohen and Havlin [2003]) property. Other studies have faced the multiscale structure of network models in a somewhat more geometric way (Boettcher and Brunson [2011]; Newman and Watts [1999]), but their findings cannot be directly applied to real-world networks.

In our geometric maps, the similarity dimension offers a reservoir of distance scales that can be used to define a geometric renormalization group for real networks (RGN) (García-Pérez, Boguñá, and Serrano [2018]). It is more convenient to work in the formulation of $\mathbb{S}^1$ model that makes explicit the similarity dimension and is mathematically more tractable. The method defines a new map by coarse-graining nodes and rescaling interactions so that longer-range connections are progressively selected at lower resolution scales. The process can be summarized as follows; see also Figure 5.

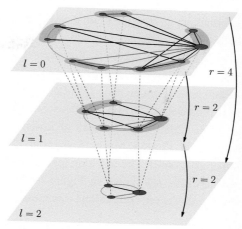

**Figure 5 Renormalization scheme.** Layer $l = 0$ represents the embedding of the original network in the $\mathbb{S}^1$ representation. Layer $l = 1, 2$ are obtained by merging pairs of consecutive nodes. Layer $l = 2$ can be obtained directly from $l = 1$ by merging groups of four nodes.

**Source:** Reprinted from García-Pérez, Boguñá, and Serrano (2018).

GEOMETRIC RENORMALIZATION GROUP

Given a real network, first obtain its geometric map using the embedding tool Mercator (García-Pérez et al. [2019]) (or an alternative method). The embedding of the network topology will assign hidden degrees $\kappa$ and angular coordinates in the similarity circle $\theta$ to every node. Then:

1. Define nonoverlapping blocks of $r$ consecutive nodes along the similarity circle and apply a coarse-graining by merging the nodes to form supernodes.

2. In the new map, place each supernode within the angular region defined by the corresponding block so that the order of nodes is preserved.

3. Connect two supernodes $i$ and $j$ in the new layer if and only if some node in block $i$ is connected to some node in block $j$ in the original network.

The operation can be iterated by taking the new layer in place of the original network to produce a multiscale unfolding. In the limit $N \to \infty$, where $N$ is the number of nodes, the RGN can be applied up to any desired scale of observation, whereas it is bounded to $\mathcal{O}(\log N)$ iterations in finite systems. In addition, notice that the transformation has the abelian semigroup structure in the sense that it is equivalent to performing two consecutive renormalization steps with $r = 2$ than one with $r = 4$.

Using this transformation, geometric scaling was found in several real scale-free networks from very different domains: the Internet at the Autonomous Systems level (Claffy et al. [2009]), the airports network (Kunegis [2013]; *OpenFlights network dataset – KONECT* [2016]), the human metabolic network at the cell level (Serrano et al. [2012]), a human protein-protein interaction network (Rolland et al. [2014]), and Drosophila Melanogaster connectome (Takemura et al. [2013]), the network of Enron emails (Klimt and Yang [2004]; Leskovec et al. [2009]), Music (Serrà et al. [2012]), and Words co-ocurrences (Milo et al. [2004]). The resulting topological features of three of the renormalized networks are shown in Figure 6. We observe that the degree distributions, degree-degree correlations – as measured by the average

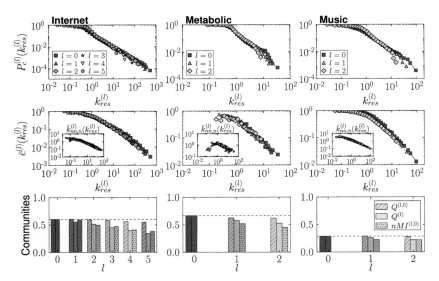

**Figure 6 Self-similarity of real networks along the RGN flow.** Each column shows the RGN flow with $r = 2$ of different topological features of the Internet AS network (left), the Human Metabolic network (middle), and the Music network (right). **Top:** Complementary cumulative distribution of rescaled degrees $k_{res}^{(l)} = k^{(l)}/\langle k^{(l)}\rangle$. **Middle:** Degree-dependent clustering coefficient over rescaled-degree classes. The inset shows the normalized average nearest neighbor degree $\bar{k}_{nn,n}(k_{res}^{(l)}) = \bar{k}_{nn}(k_{res}^{(l)})\langle k^{(l)}\rangle/\langle (k^{(l)})^2\rangle$.

**Bottom:** RGN flow of the community structure; $Q^{(l)}$ stands for the modularity in layer $l$, $Q^{(l,0)}$ is the modularity that the community structure of layer $l$ induces in the original network, and $nMI^{(l,0)}$ is the normalized mutual information between the latter and the community structure detected directly in the original network. The number of layers in each system is determined by their original size.

**Source:** Reprinted from García-Pérez, Boguñá, and Serrano (2018).

nearest neighbors degree, clustering spectra, and community structures show self-similar behavior.

The self-similarity exhibited by real-world networks can be understood in terms of their congruency with the hidden metric space network model. As we showed analytically, the $\mathbb{S}^1$ and $\mathbb{H}^2$ models are renormalizable in a geometric sense (García-Pérez, Boguñá, and Serrano [2018]). This means that if a real scale-free network is compatible with the model and admits a good embedding, the model will be able to predict its self-similarity and geometric scaling. In the $\mathbb{S}^1$ model, the transformation that ensures that the renormalized networks remain maximally congruent with the model assigns a new hidden degree $\kappa_i'$ to supernode $i$ in layer $l + 1$ as a function of the hidden degrees of the nodes it contains in the previous layer $l$ according to

$$
\kappa_i' = \left( \sum_{j=1}^{r} \kappa_j^{\beta} \right)^{1/\beta},
\tag{5.2}
$$

whereas the angular coordinate of supernode $i$, $\theta_i'$ can be taken as any (weighted) average of the angular coordinates of the nodes within the supernode. Global parameters need to be rescaled as $\mu' = \mu/r$, $\beta' = \beta$, and $R' = R/r$. With these transformations, the probability $p_{ij}'$ for two supernodes $i$ and $j$ to be connected in the new layer maintains its original form given by Eq. (2.1); see Figure 7a.

However, the preservation of the form of the connection probability shows that the $\mathbb{S}^1$ model is renormalizable but not that renormalized graphs are self-similar. For that, it is also necessary that the statistical properties of the transformed hidden degrees $\kappa'$ and angular coordinates $\theta'$ be preserved under renormalization. If the original distribution of hidden degrees is asymptotically a power law with exponent $\gamma$ and $\beta > (\gamma - 1)/2$, then the hidden degree distribution in the renormalized layers is, asymptotically, also a power law with the same exponent $\gamma$, with the only difference being the average degree. Interestingly, the global parameter controlling the clustering coefficient, $\beta$, does not change along the flow, which explains the self-similarity of the clustering spectra. Finally, the transformation for the angles preserves the ordering of nodes and the heterogeneity in their angular density, and, as a consequence, the community structure is preserved in the flow (Boguñá et al. [2010]; García-Pérez et al. [2016]; Serrano et al. [2012]; Zuev et al. [2015]); see Figure 7b. The model is, therefore, renormalizable, and RGN realizations at any scale belong to the same ensemble with a different average degree, which should be rescaled to produce self-similar replicas.

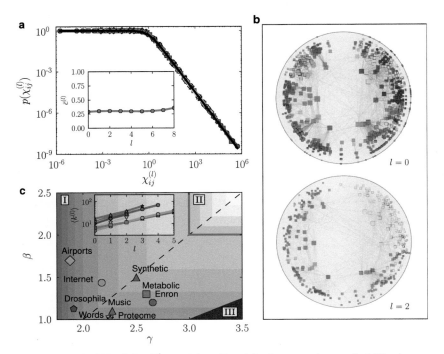

**Figure 7  RGN of the $\mathbb{S}^1$ model.** a. Empirical connection probability in a synthetic $\mathbb{S}^1$ network. Fraction of connected pairs of nodes as a function of $\chi_{ij}^{(l)} = R^{(l)} \Delta \theta_{ij}^{(l)} / (\mu^{(l)} \kappa_i^{(l)} \kappa_j^{(l)})$ in the renormalized layers, from $l = 0$ to $l = 8$, and $r = 2$. The original synthetic network has $N \sim 225{,}000$ nodes, $\gamma = 2.5$, and $\beta = 1.5$. The black dashed line shows the theoretic curve (Eq. (2.1)). The inset shows the invariance of the mean local clustering along the RGN flow. b. Hyperbolic embedding of the metabolic network (top) and its renormalized layer $l = 2$ (bottom). The colors of the nodes correspond to the community structure detected by the Louvain algorithm. Notice how the renormalized network preserves the original community structure despite being four times smaller. c. Real networks in the connectivity phase diagram. The synthetic network above also is shown. Darker blue (green) in the shaded areas represents higher values of the exponent $\nu$. The dashed line separates the $\gamma$-dominated region from the $\beta$-dominated region. In phase I, $\nu > 0$ and the network flows toward a fully connected graph. In phase II, $\nu < 0$ and the network flows toward a one-dimensional ring. The thick red line indicates $\nu = 0$ and, hence, the transition between the small-world and non–small-world phases. In region III, the degree distribution loses its scale-freeness along the flow. The inset shows the exponential increase of the average degree of the renormalized real networks $\langle k^{(l)} \rangle$ with respect to $l$.

**Source:** Reprinted from García-Pérez, Boguñá, and Serrano (2018).

**Small-World/Large-World Transition.** A good approximation of the behavior of the average degree for very large networks can be calculated by taking into account the transformation of hidden degrees in the RGN flow (Eq. (5.2)). The average degree of the renormalized layer $l+1$ as a function of the average degree in the previous layer reads

$$\langle k \rangle^{(l+1)} = r^{\nu} \langle k \rangle^{(l)}, \tag{5.3}$$

with a scaling factor $\nu$ depending on the connectivity structure of the original network. If $0 < \frac{\gamma - 1}{\beta} \leq 1$, the flow is dominated by the exponent of the degree distribution $\gamma$, and the scaling factor is given by

$$\nu = \frac{2}{\gamma - 1} - 1, \tag{5.4}$$

whereas the flow is dominated by the strength of clustering if $1 \leq \frac{\gamma - 1}{\beta} < 2$, and

$$\nu = \frac{2}{\beta} - 1. \tag{5.5}$$

Therefore, if $\gamma < 3$ or $\beta < 2$ (phase I in Figure 7c), then $\nu > 0$ and the model flows toward a highly connected graph; the average degree is preserved if $\gamma = 3$ and $\beta \geq 2$ or $\beta = 2$ and $\gamma \geq 3$, which indicates that the network is at the edge of the transition between the small-world and non–small-world phases; and $\nu < 0$ if $\gamma > 3$ and $\beta > 2$, causing the RGN flow to produce sparser networks approaching a unidimensional ring structure as a fixed point (phase II in Figure 7c). In this case, the renormalized layers eventually lose the small-world property. Finally, if $\beta < (\gamma - 1)/2$, the degree distribution becomes increasingly homogeneous as $r \to \infty$ (phase III in Fig. 7c), revealing that degree heterogeneity is present only at short scales.

In Figure 7c, several real networks are displayed in the connectivity phase diagram. All of them lie in the region of small-world networks having the fully connected network as the fixed point, which indicates that long-range connections are progressively selected by the RGN. Furthermore, all of them, except the Internet, the Airports, and the Drosophila networks, belong to the $\beta$-dominated region. The inset shows the behavior of the average degree of each layer $l$, $\langle k^{(l)} \rangle$; as predicted, it grows exponentially in all cases.

**Renormalization Explains Self-Similarity in the Brain.** Beyond the results for real networks presented above, the brain stands out as a paradigmatic system where geometric renormalization can be contrasted with multiscale empirical data. The architecture of the human brain underlies human behavior and is extremely complex with multiple scales interacting with one another.

In particular, multiscale human connectomes can be reconstructed at different length scales by hierarchical coarse-graining of anatomical regions. The multiscale organization of human connectomes at five different resolutions was explored in Zheng et al. (2020). Strikingly, the structure of the human brain remains self-similar when the resolution of observation is progressively decreased, and geometric renormalization predicts the observed multiscale properties. These results suggest that simple organizing principles underlie the multiscale architecture of human structural brain networks, where the same connectivity law dictates short- and long-range connections between different brain regions over many resolutions. The implications are varied and can be substantial for fundamental debates, such as whether the brain is working near a critical point, as well as for applications including advanced tools to simplify the digital reconstruction and simulation of the brain.

**Scaled-down Network Replicas.** The observed self-similarity of renormalized real networks and their congruency with our model can be exploited to produced scaled-down, high-fidelity replicas for useful applications; for instance, as an alternative or guidance to sampling methods in large-scale simulations and, in networked communication systems like the Internet, as a reduced test bed to analyze the performance of new routing protocols (Papadopoulos and Psounis [2007]; Papadopoulos, Psounis, and Govindan [2006]; Yao and Fahmy [2008, 2011]). Scaled-down network replicas can also be used to perform finite-size scaling of critical phenomena in real networks, so that critical exponents could be evaluated starting from a single-size instance network. Our scaled-down networks can be produced at any scale in the range in which self-similarity is preserved. The idea is to single out a specific scale after a certain number of renormalization steps and to prune the extra links to adjust its average degree to the level of the original network.

---

### SCALED-DOWN REPLICAS

Given a real network, a scaled-down network replica can be produced by applying the following algorithm:

1. Obtain a $\mathbb{S}^1$ map of the real network, including its parameters, in particular its value $\mu_0$.
2. Obtain a renormalized network layer using GRN up to the desired size and evaluate the corresponding value of $\mu_r$. Typically, $\mu_r > \mu_0$ so that the average degree of the renormalized network is larger than the original one. If this is the case, then proceed.
3. Set the auxiliary parameter $\xi = 1$.

4. Set $\mu_{new} = \xi \frac{\langle k_0 \rangle}{\langle k_r \rangle} \mu_r$.
5. Go over all links in the renormalized network and keep each existing link with probability $q_{ij} = \frac{p_{ij}(\mu_{new})}{p_{ij}(\mu_r)}$, where $p_{ij}(\mu)$ is the connection probability Eq. (2.1) with parameter $\mu$. By doing so, the probability of existence of a link in the pruned version is given by $p_{ij}(\mu_{new})$.

Finite-size effects may play an important role in real networks, so that the obtained average degree may not yet the target one. In this case, we readjust the value of $\mu_{new}$ as follows:

6. Set the tolerance for the difference between the obtained average degree after pruning and the target average degree of the original network, $\Delta$.
7. Compute the average degree $\langle k_{new} \rangle$ of the pruned layer.
8. If $\langle k_{new} \rangle - \langle k_0 \rangle > 0$, set $\xi = \xi - 0.1u$, where $u$ is a random variable uniformly distributed between $(0, 1)$. Go to step 4.
9. If $\langle k_{new} \rangle - \langle k_0 \rangle < 0$, set $\xi = \xi + 0.1u$. Go to step 4.
10. The process ends when $|\langle k_{new} \rangle - \langle k_0 \rangle| < \Delta$.

The high fidelity of scaled-down network replicas can be tested by reproducing the behavior of dynamical processes in real networks. Three different dynamical processes – the classic ferromagnetic Ising model, the susceptible-infected-susceptible (SIS) epidemic spreading model, and the Kuramoto model of synchronization – were tested in self-similar network layers of the different real networks mentioned above. Results are shown in Figure 8. Quite remarkably, for all dynamics and all networks, we observe very similar results between the original and scaled-down replicas at all scales. This is particularly interesting, as these dynamics have a strong dependence on the mesoscale structure of the underlying networks.

## 6 Navigability

Interestingly, our geometric network model offers an explanation of the efficiency of targeted transport in real networks (Boguñá et al. [2009]). Transport of information, energy, or other media through networks is a universal phenomenon in both natural and man-made complex systems. Examples include the Internet, brain, or signaling, regulatory, and metabolic networks. The information transport in these networks is not akin to diffusion. Instead, information must be delivered to specific destinations, such as specific hosts in the Internet,

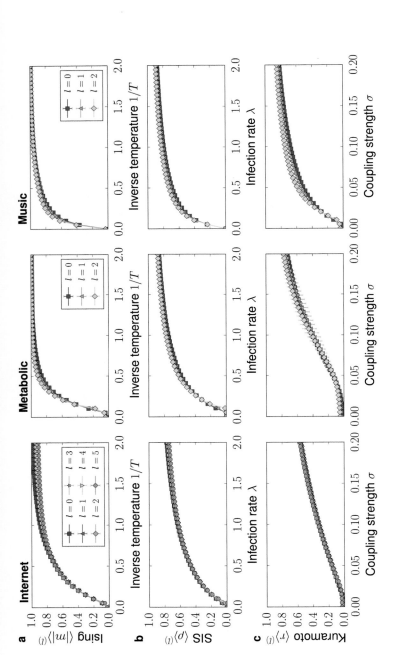

**Figure 8 Dynamics on the scaled-down replicas.** Each column shows the order parameters versus the control parameters of different dynamical processes on the original and scaled-down replicas of the Internet AS network (left), the Human Metabolic network (middle), and the Music network (right) with $r = 2$, that is, every value of $l$ identifies a network $2^l$ times smaller than the original one. All points show the results averaged over 100 simulations. Error bars indicate the standard deviations of the order parameters. a. Magnetization $\langle|m|\rangle^{(l)}$ of the Ising model as a function of the inverse temperature $1/T$. b. Prevalence $\langle\rho\rangle^{(l)}$ of the SIS model as a function of the infection rate $\lambda$. c. Coherence $\langle r\rangle^{(l)}$ of the Kuramoto model as a function of the coupling strength $\sigma$. In all cases, the curves of the smaller-scale replicas are extremely similar to the results obtained on the original networks.

Source: Reprinted from García-Pérez, Boguñá, and Serrano (2018).

groups of neurons in the brain, or genes and proteins in regulatory networks. At the same time the nodes in the network are not aware of the global network structure, so that the questions we face are whether paths to specific destinations in the network can be found without such global topology knowledge, and how optimal these paths can be. In other words, we wonder whether real complex networks are geometrically navigable.

Indeed, an underlying metric space can guide navigation in the network based on distances between nodes (Boguñá et al. [2009]). That is, instead of finding the shortest paths in the network – a computationally intensive combinatorial problem in a network that changes dynamically, such as the Internet (Korman and Peleg [2006]) – a transport process can be geometric, relying only on geodesic distances in the space. Such processes are the more efficient and robust (Muscoloni and Cannistraci [2019]) the higher the heterogeneity of the degree distribution – and so the smaller exponent $\gamma$ – and the larger the clustering coefficient – and so the larger the inverse temperature $\beta$ in our models. This defines a parameter range of navigable topologies, where most real-world networks are found (Boguñá et al. [2009]). Interestingly, networks from the $\mathbb{S}^1/\mathbb{H}^2$ model are nearly maximally efficient for such geometric navigation when the hyperbolic distance is used to define geodesic paths (Bringmann et al. [2017]; Krioukov et al. [2010]).

The explanation of this phenomenon stems from the existence, for any pair of nodes in hyperbolic networks, of topological shortest paths close to the corresponding geodesics in the underlying hyperbolic geometry. This is possible due to the existence of superhubs interconnecting all parts of the network, present as soon as $\gamma < 3$, in which case the networks are known to be ultrasmall worlds (Cohen and Havlin [2003]). As shown in Boguñá and Krioukov (2009), geometric navigation in hyperbolic networks with $\gamma < 3$ can always find these ultrashort paths, and thus navigation in these networks is asymptotically optimal. The other way around, networks that are maximally navigable by design turned out to be similar to hyperbolic networks, and many real-world networks were found to contain large fractions of their maximum-navigability skeletons (Gulyás et al. [2015]). Assuming that real-world networks evolve to have a structure efficient for their functions, these findings provide an evolutionary perspective on the emergence of latent geometry leading to structural commonalities observed in many different real-world networks.

Finally, stochastic activation–inactivation dynamics of nodes enhances the navigability of real networks with respect to the static case (Ortiz, Starnini, and Serrano [2017]). The activation dynamics may represent temporal failures of nodes due to random unknown events, or noise. Interestingly, there exists an optimal intermediate activation value, which ensures the best trade-off between the increase in the number of successful paths and a limited growth of their length. Contrary to expectations, the enhanced navigability is robust even when the most connected nodes inactivate with very high probability. In fact, it is even possible to improve the routing performance by switching on and off the hubs of the network more often than the rest of the nodes. The results indicate that some real networks are ultranavigable and remain highly navigable even if the network structure is extremely unsteady, which may have important implications for the design and evaluation of efficient routing protocols that account for the temporal or noisy nature of real complex networks.

## 7 Geometry of Weighted, Multiplex, and Growing Networks

In the previous sections, we have shown how to construct a geometric framework for complex topologies that grants practical applications rooted on the inference of hyperbolic maps. These applications include protocols for decentralized navigability that suggest a new recipe for sustainable Internet routing protocols, methods for the detection of hierarchical and community organization, and a renormalization technique that unfolds networks into a multiscale shell of self-similar layers revealing unexpected symmetries. The geometric approach to networks does not stop here, but it has been extended beyond static topologies to reach other network properties, such as weights and multiplexity, and the emergence of fundamental mechanisms from local optimization principles, such as preferential attachment.

Typically, weights in complex networks are coupled in a nontrivial way to their binary topology, playing a central role in their structural organization, function, and dynamic behavior. This is manifested, for instance, in a superlinear relation between the strength $s$ of a node (the sum of the weights attached to it) and its degree $k$ (Barrat et al. [2004]) Now, there is also empirical evidence that weights in real networks present metric properties – links involved in triangles tend to have larger normalized weights than the average link, where the normalization is given by the average weight as a function of the product of degrees of their endpoint nodes – and that the same underlying metric space ruling the network topology also shapes its weighted organization (Allard et al. [2017]).

These empirical findings are predicted by a very general model capable of reproducing the coupling with the metric space in a very simple way (Allard et al. [2017]). The model allows to fix the local properties of the nodes – their joint degree–strength distribution – while varying the coupling of the topology and, independently, of the weights with the hidden metric space. In real networks, the latter can be measured using a practical procedure that calibrates the violation of the triangle inequality under the presence of noise in the weights. Strong coupling with the metric space turns out to be a highly plausible explanation for the observed weighted organization of many real networks, including metabolic networks, human brain connectomes, international trade, cargo ship movements, and more, although in some systems – for example the US airports network – the coupling of weights with the underlying geometry is weak in contrast with topology, which suggests that the formation of connections and the assignment of their magnitude might be ruled by different processes.

Apart from weights, interactions between pairs of nodes in real networks can be of different types. This leads to multiplex representations (Bianconi [2018]), where links of different nature coexist and form different layers. These structures are not random combinations of single networks but, in contrast, exhibit significant hidden geometric correlations (Kleineberg et al. [2016]). In real multiplexes, coordinates of nodes in hyperbolic maps of each separate layer are significantly correlated, and so distances between nodes in the corresponding underlying hyperbolic spaces are significantly correlated. These correlations are found in real multiplexes – the Internet, protein interaction networks, collaboration networks, and more – at the level of both radial and angular coordinates.

Radial correlations measured in real multiplexes are equivalent to correlations among node degrees (Min et al. [2014]; Nicosia and Latora [2015]; Serrano, Buzna, and Boguñá [2015]). On the other hand, the observed correlation among the angular similarity coordinates is a genuine geometric feature with important theoretical and practical implications. Specifically, hidden interlayer geometric correlations mitigate the vulnerability of multiplexes to targeted attacks on high-degree nodes, making multiplex networks unexpectedly robust (Kleineberg et al. [2017]). Without geometric correlations, multiplexes exhibit an abrupt breakdown of mutual connectivity, even with interlayer degree correlations. With geometric correlations, a multistep cascading process is instead observed, which does not destroy the system completely but leads into an eventually smooth percolation transition, with results suggesting that it can be fully continuous in the thermodynamic limit.

On the practical side, geometric correlations can facilitate the definition and detection of multidimensional communities, which are sets of nodes that are simultaneously similar in multiple layers, and enable accurate trans-layer link prediction, meaning that connections in one layer can be predicted by observing the hidden geometric space of another layer. Also, sufficiently strong geometric correlations allow efficient targeted navigation in the multilayer system using only local knowledge, outperforming navigation in the single layers. These effects of geometric correlations can be assessed using a geometric multiplex network model that generates multiplex layers with realistic synthetic topologies where correlations – both radial and angular – can be tuned without altering the topological characteristics of each individual layer (Kleineberg et al. [2017]).

Finally, beyond the structure of static networks, the geometric approach can also explain mechanisms that underlie network growth in terms of local optimization processes such as preferential attachment, a common explanation for the emergence of scaling in growing networks (Barabási and Albert [1999]). If new connections are made preferentially to more popular nodes, then the resulting distribution of the number of connections possessed by nodes follows power laws, as observed in many real networks (Newman [2010]). Preferential attachment can be a consequence of different underlying processes based on node fitness, ranking, optimization, random walks, or duplication (Caldarelli et al. [2002]; Dorogovtsev, Mendes, and Samukhin [2001]; D'Souza et al. [2007]; Fortunato, Flammini, and Menczer [2006]; Pastor-Satorras, Smith, and Sole [2003]), but it can also emerge from a geometric description in which new connections optimize the product between popularity and similarity. This idea has been formalized in the PSO model (Papadopoulos et al. [2012]) that can be thought of as a generalization of our geometric $\mathbb{S}^1/\mathbb{H}^2$ model to growing networks where the latent space is not hyperbolic but de Sitter space $d\mathbb{S}^{1,D}$, which has the same Lorentz group $SO(1, D + 1)$ of symmetries (Krioukov et al. [2012]). As opposed to preferential attachment, PSO describes with good accuracy the large-scale evolution of technological (the Internet), social (trust relationships between people), and biological (metabolic) networks that grow through the sequential addition of new nodes that connect to older ones in the graph, predicting the probability of new links with high precision.

However, many real systems evolve in a self-similar way that preserves their topology throughout the growth process over a long time span that is better explained by branching of fundamental units – whether those be scientific fields or countries (Zheng et al. [2021]). The Geometric Branching Growth model predicts this evolution and explains the symmetries observed (Zheng et al. [2021]). The model produces multiscale unfolding of a network in a sequence of scaled-up replicas preserving network features, including clustering and community structure. When combined with scaled-down network replicas produced by geometric renormalization (García-Pérez, Boguñá, and Serrano [2018]), the model provides a full up-and-down self-similar multiscale unfolding of complex networks that covers both large and small scales. Practical applications in real instances that require optimization or control of system size in complex networks are countless. They include the tuning of network size for best response to external influence and finite-size scaling to assess critical behavior under random link failures, assessment of scalability issues in dynamic processes in core functions of real networks, such as in Internet routing protocols, and more.

## 8 Conclusions

It is a very well-established empirical fact that most real complex networks share a very special set of universal features. Among the most relevant ones, networks have heterogeneous degree distributions, are small-worlds, and have strong clustering. Our hidden metric space network model (Krioukov et al. [2010]; Papadopoulos et al. [2012]; Serrano et al. [2008]), independently of its $\mathbb{S}^1$ or $\mathbb{H}^2$ formulation, provides a very natural explanation of these properties with a limited number of parameters and using an effective hyperbolic geometry of dimension 2. Even if the model and the corresponding renormalization group can be formulated in $D$ dimensions, the high clustering coefficient observed in real networks poses an upper limit on the potential dimension of the similarity space (Dall and Christensen [2002]; García-Pérez, Boguñá, and Serrano [2018]). Besides, in hyperbolic geometry space expands exponentially fast,[5] even in $D = 2$, so that networks can be faithfully embedded in the one-dimensional $\mathbb{S}^1$ model even if the original network has a higher-dimensional similarity subspace. This is in line with the accumulated empirical evidence, which unambiguously supports the one-dimensional similarity plus degrees as an extremely good proxy for the geometry of real networks.

As we have discussed in Section 3, our models can be used both as topology generators and as a mapping tool to obtain geometric representations of real-world networks in the hyperbolic plane. One of the obvious advantages offered by geometric maps of real networks is that many methods that have been designed for extended systems in physics or for high-dimensional data in computer science can be adapted to networks. Here we have introduced three such applications: community detection, geometric navigation, and geometric renormalization of complex networks. Concerning the latter, the existence of a metric space underlying complex networks allows us to define a geometric renormalization group that reveals their multiscale nature. Quite strikingly, our geometric models of scale-free networks are shown to be self-similar under such renormalization. Even more important is the finding that self-similarity under geometric renormalization is a ubiquitous symmetry of real-world scale-free networks, which provides new evidence supporting the hypothesis that hidden metric spaces do underlie real networks.

---

[5] Asymptotically, the volume of a ball of radius $r$ grows as $V(r) \sim e^{(D-1)r}$.

The geometric renormalization group presented in this work is similar in spirit to topological renormalization (studied in Goh et al. [2006]; Kim et al. [2007]; Radicchi et al. [2008]; Rozenfeld et al. [2010]; Song et al. [2005, 2006]). However, instead of using shortest paths as a source of length scales to explore the fractality of networks, geometric renormalization uses a continuum geometric framework that allows us to unveil the role of degree heterogeneity and clustering in the self-similarity properties of networks. In our model, a crucial point is the explicit contribution of degrees to the probability of connection, giving the clue by which we can produce both short-range and long-range connections using a single mechanism captured in a universal connectivity law. The combination of similarity with degrees is a necessary condition to make the model predictive of the multiscale properties of real networks.

From a fundamental point of view, the geometric renormalization group introduced here has proven to be an exceptional tool to unravel the global organization of complex networks across scales and promises to become a standard methodology to analyze real complex networks. It can also help in areas like the study of metapopulation models, in which transportation fluxes or population movements happen on both local and global scales (Colizza, Pastor-Satorras, and Vespignani [2007]). From a practical point of view, we envision many applications. In large-scale simulations, scaled-down network replicas could serve as an alternative or guidance to sampling methods, or for fast-track exploration of rough parameter spaces in the search of relevant regions. Scaled-down versions of real networks could also be applied to perform finite-size scaling, which would allow for the determination of critical exponents from single snapshots of their topology. Other possibilities include the development of a new multilevel community detection method (Abou-Rjeili and Karypis [2006]; Karypis and Kumar [1999]) that would exploit the mesoscopic information encoded in the different observation scales.

Our models have been extended to weighted networks, multiplexes, and growing networks, but they still have to be extended to systems with asymmetric interactions, represented as directed networks. Examples are found in many different domains, from cellular networks, like in metabolic, gene-regulatory, or neural networks; technological systems, like the World Wide Web or the Internet, to social systems, as friendship interaction between two persons can be perceived as different from person A to person B or from B to A. Such asymmetry implies that the geometric paradigm must be adapted to this particular type of system. The main caveat in this case stems from the breakdown of the triangle inequality in directed networks, a fundamental property in any metric space. Yet, given the ubiquitous presence of asymmetric relationships in real systems, such extension promises to represent an important step forward in our understanding of complex systems.

Network geometry today represents the best description of real complex networks, providing new insights into the fundamental principles that shape their structure. Geometric models are able to encode in simple connectivity laws most of the complex topological patterns of interactions in real complex systems. Future advances need to amplify the geometric framework to include the study of dynamical processes that act as the necessary bridge between network structure and function.

# References

Abou-Rjeili, A., & Karypis, G. (2006). Multilevel algorithms for partitioning power-law graphs. In *Proceedings 20th IEEE International Parallel & Distributed Processing Symposium*. doi: 10.1109/IPDPS.2006.1639360.

Alanis-Lobato, G., Mier, P., & Andrade-Navarro, M. A. (2016a). Efficient embedding of complex networks to hyperbolic space via their Laplacian. *Sci. Rep.*, *6*, 30108.

Alanis-Lobato, G., Mier, P., & Andrade-Navarro, M. A. (2016b, Nov. 15). Manifold learning and maximum likelihood estimation for hyperbolic network embedding. *Applied Network Science*, *1*(1), 10. doi: https://doi.org/10.1007/s41109-016-0013-0

Allard, A., & Serrano, M. Á. (2020). Navigable maps of structural brain networks across species. *PLOS Computational Biology*, *16*(2), e1007584.

Allard, A., Serrano, M. Á., García-Pérez, G., & Boguñá, M. (2017). The geometric nature of weights in real complex networks. *Nat. Commun.*, *8*, 14103.

Alvarez-Hamelin, J. I., Dall'Asta, L., Barrat, A., & Vespignani, A. (2008). K-core decomposition of internet graphs: hierarchies, selfsimilarity and measurement biases. *Networks and Heterogeneous Media*, *3*(2), 371–393.

Amaral, L. A. N. (2008). A truer measure of our ignorance. *Proc. Natl. Acad. Sci. USA*, *105*(19), 6795–6796.

Amaral, L. A. N., Scala, A., Barthélemy, M., & Stanley, H. E. (2000). Classes of small-world networks. *Proc. Natl. Acad. Sci. USA*, *97*(21), 11149–11152.

Barabási, A. L., & Albert, R. (1999). Emergence of scaling in random networks. *Science*, *286*(5439), 509–512.

Barrat, A., Barthélemy, M., Pastor-Satorras, R., & Vespignani, A. (2004). The architecture of complex weighted networks. *Proc. Natl. Acad. Sci. USA*, *101*(11), 3747–3752.

Barrat, A., Barthélemy, M., & Vespignani, A. (2008). *Dynamical processes on complex networks*. Cambridge: Cambridge University Press.

Barthélemy, M. (2011). Spatial networks. *Phys. Rep.*, *499*(1–3), 1–101.

Bianconi, G. (2018). *Multilayer networks: structure and function*. Oxford: Oxford University Press.

Blasius, T., Friedrich, T., Krohmer, A. et al. (2018, Apr.). Efficient embedding of scale-free graphs in the hyperbolic plane. *IEEE/ACM Trans. Netw.*, *26*(2), 920–933. doi: https://doi.org/10.1109/TNET.2018.2810186

Blondel, V. D., Guillaume, J.-L., Lambiotte, R., & Lefebvre, É. (2008). Fast unfolding of communities in large networks. *J. Stat. Mech.*, *2008*(10), P10008.

Boettcher, S., & Brunson, C. (2011). Renormalization group for critical phenomena in complex networks. *Frontiers in Physiology*, *2*, 102. doi: https://doi.org/10.3389/fphys.2011.00102

Boguñá, M., Bonamassa, I., Domenico, M. D. et al. (2020). Network geometry. *Nat Rev Phys* **3**, 114–135. https://doi.org/10.1038/s42254-020-00264-4

Boguñá, M., & Krioukov, D. (2009). Navigating ultrasmall worlds in ultrashort time. *Phys. Rev. Lett.*, *102*(058701). (arXiv:0809.2995v1)

Boguñá, M., Krioukov, D., Almagro, P., & Serrano, M. Á. (2020, Apr.). Small worlds and clustering in spatial networks. *Phys. Rev. Research*, *2*, 023040. doi: https://doi.org/10.1103/PhysRevResearch.2.023040

Boguñá, M., Krioukov, D., & Claffy, K. (2009). Navigability of complex networks. *Nat. Phys.*, *5*(1), 74–80.

Boguñá, M., Papadopoulos, F., & Krioukov, D. (2010). Sustaining the Internet with hyperbolic mapping. *Nat. Commun.*, *1*, 62. doi: https://doi.org/10.1038/ncomms1063

Boguñá, M., Pastor-Satorras, R., & Vespignani, A. (2004). Cut-offs and finite size effects in scale-free networks. *Eur. Phys. J. B*, *38*(2), 205–209.

Bringmann, K., Keusch, R., Lengler, J., Maus, Y., & Molla, A. R. (2017). Greedy routing and the algorithmic small-world phenomenon. In *PODC '17: Proceedings of the ACM Symposium on Principles of Distributed Computing*. doi: https://doi.org/10.1145/3087801.3087829

Caldarelli, G., Capocci, A., De Los Rios, P., & Muñoz, M. A. (2002, December). Scale-free networks from varying vertex intrinsic fitness. *Phys. Rev. Lett.*, *89*(25), 258702. doi: https://doi.org/10.1103/PhysRevLett.89.258702

Claffy, K., Hyun, Y., Keys, K., Fomenkov, M., & Krioukov, D. (2009). Internet mapping: From art to science. In *2009 Cybersecurity Applications Technology Conference for Homeland Security* (pp. 205–211). New York: IEEE. doi: https://doi.org/10.1109/CATCH.2009.38

Cohen, R., & Havlin, S. (2003). Scale-free networks are ultrasmall. *Phys. Rev. Lett.*, *90*(5), 058701.

Colizza, V., Pastor-Satorras, R., & Vespignani, A. (2007). Reaction-diffusion processes and metapopulation models in heterogeneous networks. *Nat. Phys.*, *3*, 276–282.

Dall, J., & Christensen, M. (2002). Random geometric graphs. *Phys. Rev. E*, *66*(1), 016121. doi: https://doi.org/10.1103/PhysRevE.66.016121

Dorogovtsev, S. N., & Mendes, J. F. F. (2003). *Evolution of networks: From biological nets to the Internet and WWW*. Oxford: Oxford University Press.

Dorogovtsev, S. N., Mendes, J. F. F., & Samukhin, A. N. (2001). A. size-dependent degree distribution of a scale-free growing network. *Phys. Rev. E*, *63*(6), 062101.

D'Souza, R., Borgs, C., Chayes, J., Berger, N., & Kleinberg, R. (2007). Emergence of tempered preferential attachment from optimization. *PNAS*, *104*(15), 6112–6117.

Fortunato, S., Flammini, A., & Menczer, F. (2006). Scale-free network growth by ranking. *Phys. Rev. Lett.*, *96*(21), 218701.

García-Pérez, G., Allard, A., Serrano, M. Á., & Boguñá, M. (2019, Dec.). Mercator: uncovering faithful hyperbolic embeddings of complex networks. *New Journal of Physics*, *21*(12), 123033. doi: https://doi.org/10.1088%2F1367-2630%2Fab57d2

García-Pérez, G., Boguñá, M., Allard, A., & Serrano, M. Á. (2016, Sept.). The hidden hyperbolic geometry of international trade: World Trade Atlas 1870–2013. *Sci. Rep.*, *6*, 33441. doi: https://doi.org/10.1038/srep33441

García-Pérez, G., Boguñá, M., & Serrano, M. Á. (2018). Multiscale unfolding of real networks by geometric renormalization. *Nat. Phys.*, *14*(6), 583–589. doi: https://doi.org/10.1038/s41567-018-0072-5

García-Pérez, G., Serrano, M. Á., & Boguñá, M. (2018). Soft communities in similarity space. *J. Stat. Phys.*, *173*(3–4), 775–782. doi: https://doi.org/10.1007/s10955-018-2084-z

Goh, K. I., Salvi, G., Kahng, B., & Kim, D. (2006). Skeleton and fractal scaling in complex networks. *Phys. Rev. Lett.*, *96*, 018701.

Guimerà, R., Mossa, S., Turtschi, A., & Amaral, L. A. N. (2005). The worldwide air transportation network: Anomalous centrality, community structure, and cities. *Proc. Natl. Acad. Sci. USA*, *10*(22), 7794–7799.

Gulyás, A., Bíró, J. J., Kőrösi, A., Rétvári, G., & Krioukov, D. (2015). Navigable networks as Nash equilibria of navigation games. *Nat. Commun.*, *6*(1), 7651. doi: https://doi.org/10.1038/ncomms8651

Kadanoff, Leo P. (2000). *Statistical physics: Statics, dynamics and renormalization*. Singapore: World Scientific.

Karypis, G., & Kumar, V. (1999). A fast and highly quality multilevel scheme for partitioning irregular graphs. *SIAM Journal on Scientific Computing*, *20*(1), 359–392.

Kim, J. S., Goh, K. I., Hahng, B., & Kim, D. (2007). Fractality and self-similarity in scale-free networks. *New J. Phys.*, *9*, 177.

Kleineberg, K.-K., Boguñá, M., Serrano, M. Á., & Papadopoulos, F. (2016). Hidden geometric correlations in real multiplex networks. *Nat. Phys.*, *12*(11), 1076–1081.

Kleineberg, K.-K., Buzna, L., Papadopoulos, F., Boguñá, M., & Serrano, M. Á. (2017). Geometric correlations mitigate the extreme vulnerability of multiplex networks against targeted attacks. *Phys. Rev. Lett.*, *118*, 218301.

Klimt, B., & Yang, Y. (2004). Introducing the Enron Corpus. In *CEAS 2004 – First Conference on Email and Anti-Spam*, July 30–31, 2004, Mountain View, CA. Accessed at http://dblp.uni-trier.de/db/conf/ceas/ceas 2004.html#KlimtY04

Korman, A., & Peleg, D. (2006). Dynamic routing schemes for general graphs. In M. Bugliesi, B. Preneel, V. Sassone, & I. Wegener (eds.), *ICALP: Proceedings of Int. Colloquium on Automata, Languages and Programming* (vol. 4051, pp. 619–630). Springer. doi: https://doi.org/10.1007/11786986_54

Krioukov, D., Kitsak, M., Sinkovits et al. (2012). Network cosmology. *Sci. Rep.*, *2*. Accessed at doi: https://doi.org/10.1038/srep00793

Krioukov, D., Papadopoulos, F., Kitsak, M., Vahdat, A., & Boguñá, M. (2010). Hyperbolic geometry of complex networks. *Phys. Rev. E*, *82*(3), 036106.

Krioukov, D., Papadopoulos, F., Vahdat, A., & Boguñá, M. (2009, Sep.). Curvature and temperature of complex networks. *Phys. Rev. E*, *80*(3), 035101. Accessed at http://link.aps.org/doi/10.1103/PhysRevE.80.035101 doi: https://doi.org/10.1103/PhysRevE.80.035101

Kunegis, J. (2013). KONECT – The Koblenz Network Collection. In *Proceedings of the International Conference on World Wide Web Companion* (pp. 1343–1350). Accessed at http://konect.cc/

Leskovec, J., Lang, K. J., Dasgupta, A., & Mahoney, M. W. (2009). Community structure in large networks: Natural cluster sizes and the absence of large well-defined clusters. *Internet Mathematics*, *6*(1), 29–123. doi: https:doi.org/10.1080/15427951.2009.10129177

Maldacena, J. (1998). The large $N$ limit of superconformal field theories and supergravity. *Adv. Theor. Math. Phys.*, *2*(2), 231–252. doi: https://doi.org/10.4310/ATMP.1998.v2.n2.a1

Mandelbrot, B. (1961). On the theory of word frequencies and on related Markovian models of discourse. In *Proceedings of the Twelve Symposia in Applied Mathematics, Roman Jakobson Editor. Structure of Language and Its Mathematical Aspects, New York, USA* (pp. 190–219). Providence, RI: American Mathematical Society.

Milo, R., Itzkovitz, S., Kashtan, N. et al. (2004, March). Superfamilies of evolved and designed networks. *Science*, *303*(5663), 1538–1542.

Min, B., Yi, S. D., Lee, K.-M., & Goh, K.-I. (2014). Network robustness of multiplex networks with interlayer degree correlations. *Phys. Rev. E*, *89*(4), 042811.

Muscoloni, A., & Cannistraci, C. V. (2018). A nonuniform popularity-similarity optimization (nPSO) model to efficiently generate realistic complex networks with communities. *New J. Phys.*, *20*(5), 052002. doi: https://doi.org/10.1088/1367-2630/aac06f

Muscoloni, A., & Cannistraci, C. V. (2019). Navigability evaluation of complex networks by greedy routing efficiency. *Proc. Natl. Acad. Sci.*, *116*(5), 1468–1469. doi: https://doi.org/10.1073/pnas.1817880116

Muscoloni, A., Thomas, J. M., Ciucci, S., Bianconi, G., & Cannistraci, C. V. (2017). Machine learning meets complex networks via coalescent embedding in the hyperbolic space. *Nature Communications*, *8*(1), 1615. https://doi.org/10.1038/s41467-017-01825-5

Newman, M., & Watts, D. (1999). Renormalization group analysis of the small-world network model. *Physics Letters A*, *263*(4–6), 341–346. Accessed at www.sciencedirect.com/science/article/pii/S037596019 9007574 doi: https://doi.org/10.1016/S0375-9601(99)00757-4

Newman, M. E. J. (2010). *Networks: An introduction.* Oxford: Oxford University Press.

Newman, M. E. J., & Girvan, M. (2004, Feb.). Finding and evaluating community structure in networks. *Phys. Rev. E*, *69*(2), 026113. doi: https://doi.org/10.1103/PhysRevE.69.026113

Nicosia, V., & Latora, V. (2015). Measuring and modeling correlations in multiplex networks. *Phys. Rev. E*, *92*, 032805.

*Openflights network dataset – KONECT.* (2016, Sept.). Accessed at http://konect.uni-koblenz.de/networks/openflights

Ortiz, E., Starnini, M., & Serrano, M. Á. (2017). Navigability of temporal networks in hyperbolic space. *Sci. Rep.*, *7*, 15054.

Papadopoulos, F., Aldecoa, R., & Krioukov, D. (2015, Aug.). Network geometry inference using common neighbors. *Phys. Rev. E*, *92*(2). 022807. doi: https://doi.org/10.1103/PhysRevE.92.022807

Papadopoulos, F., Kitsak, M., Serrano, M. Á., Boguñá, M., & Krioukov, D. (2012). Popularity versus similarity in growing networks. *Nature*, *489*(7417), 537–540. doi: https://doi.org/10.1038/nature 11459

Papadopoulos, F., Krioukov, D., Boguñá, M., & Vahdat, A. (2010). Greedy forwarding in dynamic scale-free networks embedded in hyperbolic metric spaces. In *2010 Proceedings IEEE Infocom* (pp. 1–9).

Papadopoulos, F., Psomas, C., & Krioukov, D. (2015). Network mapping by replaying hyperbolic growth. *IEEE/ACM Trans. Netw.*, *23*(1), 198–211. doi: https://doi.org/10.1109/TNET.2013.2294052

Papadopoulos, F., & Psounis, K. (2007, Oct.). Efficient identification of uncongested internet links for topology downscaling. *SIG-COMM Comput. Commun. Rev.*, *37*(5), 39–52. doi: https://doi.org/10.1145/1290168. 1290173

Papadopoulos, F., Psounis, K., & Govindan, R. (2006, Dec.). Performance preserving topological downscaling of internet-like networks. *IEEE Journal on Selected Areas in Communications*, *24*(12), 2313–2326. doi: https://doi.org/10.1109/JSAC.2006.884029

Pastor-Satorras, R., Smith, E., & Sole, R. V. (2003). Evolving protein interaction networks through gene duplication. *J. Theor. Biol.*, *222*(2), 199–210.

Radicchi, F., Ramasco, J. J., Barrat, A., & Fortunato, S. (2008, Oct.). Complex networks renormalization: Flows and fixed points. *Phys. Rev. Lett.*, *101*(14), 148701. doi: https://doi.org/10.1103/PhysRevLett.101.148701

Rolland, T., Taşan, M., Charloteaux, B. et al. (2014). A proteome-scale map of the human interactome network. *Cell*, *159*(5), 1212–1226. Accessed at www.sciencedirect.com/science/article/pii/S0092867414014226 doi: https: //doi.org/10.1016/j.cell.2014.10.050

Rozenfeld, H. D., Song, C., & Makse, H. A. (2010, Jan.). Small-world to fractal transition in complex networks: A renormalization group approach. *Phys. Rev. Lett.*, *104*(2), 025701. doi: https://doi.org/10.1103 /PhysRevLett.104.025701

Sarveniazi, A. (2014). An actual survey of dimensionality reduction. *American Journal of Computational Mathematics*, *4*(2), 55–72.

Serrà, J., Corral, A., Boguñá, M., Haro, M., & Arcos, J. L. (2012). Measuring the evolution of contemporary western popular music. *Sci. Rep.*, *2*. Accessed at www.nature.com/srep/2012/120726/srep00521/full/srep00521.html doi: https://doi.org/10.1038/srep00521

Serrano, M. Á., Boguñá, M., & Sagues, F. (2012). Uncovering the hidden geometry behind metabolic networks. *Mol. BioSyst.*, *8*(3), 843–850. doi: https://doi.org/10.1039/C2MB05306C

Serrano, M. Á., Buzna, L., & Boguñá, M. (2015). Escaping the avalanche collapse in self-similar multiplexes. *New J. Phys.*, *17*, 053033.

Serrano, M. Á., Krioukov, D., & Boguñá, M. (2008). Self-similarity of complex networks and hidden metric spaces. *Phys. Rev. Lett.*, *100*(7), 078701.

Serrano, M. Á., Krioukov, D., & Boguñá, M. (2011, Jan.). Percolation in self-similar networks. *Phys. Rev. Lett.*, *106*(4), 048701. doi: https://doi.org/10.1103/PhysRevLett.106.048701

Song, C., Havlin, S., & Makse, H. A. (2005). Self-similarity of complex networks. *Nature*, *433*(7024), 392–395.

Song, C., Havlin, S., & Makse, H. A. (2006). Origins of fractality in the growth of complex networks. *Nat. Phys.*, *2*(4), 275–281.

Stanley, H. E. (1971). *Introduction to phase transitions and critical phenomena.* Oxford: Oxford University Press.

Starnini, M., Ortiz, E., & Serrano, M. Á. (2019). Geometric randomization of real networks with prescribed degree sequence. *New J. Phys., 21*(5), 053039.

Takemura, S.-y., Bharioke, A., Lu, Z. et al. (2013). A visual motion detection circuit suggested by drosophila connectomics. *Nature, 500*(7461), 175–181.

Watts, D. J., & Strogatz, S. H. (1998). Collective dynamics of "small-world" networks. *Nature, 393*(6684), 440–442.

Wilson, K. G. (1975). The renormalization group: Critical phenomena and the kondo problem. *Rev. Mod. Phys., 47*(4), 773–840.

Wilson, K. G. (1983, Jul.). The renormalization group and critical phenomena. *Rev. Mod. Phys., 55*(3), 583–600. doi: https://doi.org/10.1103/RevModPhys.55.583

Yao, W. M., & Fahmy, S. (2008, Apr.). Downscaling network scenarios with Denial of Service (DoS) attacks. In *2008 IEEE Sarnoff Symposium* (pp. 1–6). doi: https://doi.org/10.1109/SARNOF.2008.4520099

Yao, W. M., & Fahmy, S. (2011, June). Partitioning network testbed experiments. In *2011 31st International Conference on Distributed Computing Systems* (pp. 299–309). doi: https://doi.org/10.1109/ICDCS.2011.22

Zheng, M., Allard, A., Hagmann, P., Alemán-Gómez, Y., & Serrano, M. Á. (2020). Geometric renormalization unravels self-similarity of the multiscale human connectome. *Proc. Natl. Acad. Sci., 117*(33), 20244–20253. doi: https://doi.org/10.1073/pnas.1922248117

Zheng, M., García-Pérez, G., Boguñá, & M., Serrano, M. Á. (2021). Scaling-up real networks by geometric branching growth. *Proc. Natl. Acad. Sci., 118*(21), e2018994118. doi: https://doi.org/10.1073/pnas.2018994118

Zuev, K., Boguñá, M., Bianconi, G., & Krioukov, D. (2015). Emergence of soft communities from geometric preferential attachment. *Sci. Rep., 5*, 9421. doi: https://doi.org/10.1038/srep09421

# Acknowledgments

We acknowledge support from a James S. McDonnell Foundation Scholar Award in Complex Systems; the ICREA Academia award, funded by the *Generalitat de Catalunya*; *Agencia estatal de investigación* project no. PID2019-106290GB-C22/AEI/10.13039/501100011033; the Spanish *Ministerio de Ciencia, Innovación y Universidades* project no. FIS2016-76830-C2-2-P (AEI/FEDER, UE); the project *Mapping Big Data Systems: embedding large complex networks in low-dimensional hidden metric spaces, Ayudas Fundación BBVA a Equipos de Investigación Científica 2017*; and *Generalitat de Catalunya* grant no. 2017SGR1064.

**Cambridge Elements** ≣

# The Structure and Dynamics of Complex Networks

## Guido Caldarelli

*Ca' Foscari University of Venice*

Guido Caldarelli is Full Professor of Theoretical Physics at Ca' Foscari University of Venice. Guido Caldarelli received his Ph.D. from SISSA, after which he held postdoctoral positions in the Department of Physics and School of Biology, University of Manchester, and the Theory of Condensed Matter Group, University of Cambridge. He also spent some time at the University of Fribourg in Switzerland, at École Normale Supérieure in Paris, and at the University of Barcelona. His main research focus is the study of networks, mostly analysis and modelling, with applications from financial networks to social systems as in the case of disinformation. He is the author of more than 200 journal publications on the subject, and three books, and is the current President of the Complex Systems Society (2018 to 2021).

### About the Series

This cutting-edge new series provides authoritative and detailed coverage of the underlying theory of complex networks, specifically their structure and dynamical properties. Each Element within the series will focus upon one of three primary topics: static networks, dynamical networks, and numerical/computing network resources.

**Cambridge Elements** ☰

# The Structure and Dynamics of Complex Networks

Printed in the United States
by Baker & Taylor Publisher Services